A
NATURALIST IN THE GRAN CHACO

A NATURALIST IN THE GRAN CHACO

BY

SIR JOHN GRAHAM KERR, F.R.S.

Honorary Fellow of Christ's College, Cambridge
and sometime Regius Professor of Zoology
in the University of Glasgow

CAMBRIDGE
AT THE UNIVERSITY PRESS
1950

CAMBRIDGE
UNIVERSITY PRESS

University Printing House, Cambridge CB2 8BS, United Kingdom

Cambridge University Press is part of the University of Cambridge.

It furthers the University's mission by disseminating knowledge in the pursuit of education, learning and research at the highest international levels of excellence.

www.cambridge.org
Information on this title: www.cambridge.org/9781107495050

First published 1950
First paperback edition 2015

A catalogue record for this publication is available from the British Library

ISBN 978-1-107-49505-0 Paperback

PREFACE

During recent troublous years I have found much comfort in re-reading my diaries written long ago in the Gran Chaco and thereby escaping for the time being into what seemed a different world. I am not without hope that others too may find such comfort even though of lesser degree in the perusal of the excerpts from my Chaco diaries printed in this volume.

Their main appeal will of course be to naturalists who will find therein recorded many observations which though made long ago have not hitherto been published. In Part II dealing with the South American lungfish, that strange link between water-breathing fish and air-breathing land animals, they will find a short account of my endeavour to reach what I regard as the ideal of the scientific naturalist—the preliminary study of the living animal in its natural environment, followed by the more precise and intensive investigations of the laboratory—in other words the intimate linking together of field natural history and laboratory research.

The student of sociology or anthropology will be more particularly interested in my picture of the Natokoi or Toba Indians of the Pilcomayo, who, isolated to a great extent from outside influences through being surrounded by hostile tribes, had managed to linger on in what must be regarded as an extraordinarily primitive stage of communal evolution. The accuracy of my picture has been greatly enhanced by two factors: (1) the common interest we had as observers and hunters of wild animals, and (2) the fact that I was not among them in order to dispose of my own wares—material or spiritual. The trader and the missionary alike are apt to be prevented from getting into the most intimate contact with primitive peoples by an impenetrable barrier of suspicion. They are apt to be hampered too by preconceived ideas—religious, scientific or what not—which tend to mislead both in matters of observation and in the interpretation of what is observed.

PREFACE

In considering the make-up of this volume I have decided, after due consideration,

(1) to make no attempt to provide detailed maps: Chaco surveys are hampered by the impermanence of the only available landmarks such as the courses of rivers and the boundaries of swamp and forest;

(2) as regards nomenclature to use as a rule the scientific names of animals and plants in vogue at the date of my expedition; and

(3) in the matter of spelling Spanish words to follow South American custom, e.g. Bermejo, Biscacha, Vaqueano rather than Vermejo, Viscacha, Baqueano.

I have to make grateful acknowledgment of assistance given by many friends and colleagues regarding various matters of detail: Mr R. H. Tottenham, H.B.M. Chargé d'Affaires at Asunción; the Directors of the British Museum (Natural History), the La Plata Museum, the Buenos Aires Museum of Natural Sciences, Messrs D. A. Bannerman, R. B. Benson, Max Birabín, W. S. Bristowe, H. A. Gleason, M. Hinton, A. J. Montague, of the British Council, and H. F. Schwarz, of the American Museum of Natural History.

My comrades on the Pilcomayo Expedition, Messrs Andrew Pride and R. J. Hunt of the South American Missionary Society, and my many Indian friends and helpers, are no longer accessible to thanks, but it is only right to record the warm and grateful memories they have left behind.

I am indebted to my wife for much valuable help and criticism and to my son Ronald for a number of drawings of native implements.

JOHN GRAHAM KERR

21 October 1949

CONTENTS

LIST OF PLATES

(The plates are bound together at the end of the book, between pages 232 and 233).

LIST OF PLATES

MAPS

Available for download from www.cambridge.org/9781107495050

Part I

PILCOMAYO EXPEDITION

1889–91

CHAPTER I

PROLOGUE

It was on a wintry afternoon in February 1889 that I, a young medical student of nineteen, returning home from my classes in the University of Edinburgh, picked up at the book-stall in Waverley station a number of *Nature* which determined the whole future of my life: for therein I read the following letter:

Opportunity for a Naturalist

Captain Juan Page, of the Argentine Navy, who is now in London, and read a paper on the exploration of the Rio Vermejo and Rio Pilcomayo at the last meeting of the Royal Geographical Society, has undertaken a new expedition for the survey of the Pilcomayo from the Paraná to the frontiers of Bolivia. Captain Page would be glad to give a place on the Staff of this Expedition to a naturalist, who would have the opportunity of investigating the almost unknown flora and fauna of the Gran Chaco, through which the Pilcomayo runs. The Expedition will start from Buenos Ayres in June next, and be absent about six months. The naturalist would have to find his passage out to Buenos Ayres and home, and his own equipment and collecting materials, but on joining the Expedition would be free from charges. I should be glad to put any qualified person who might wish to avail himself of this excellent opportunity of exploring a most interesting country in communication with Captain Page.

P. L. SCLATER

Zoological Society of London,
3, Hanover Square, London, W.
February 4.

I had been an enthusiastic reader of books on exploration and natural history—Waterton's *Wanderings in South America*, Darwin's *Voyage of the Beagle*, Bates's *Naturalist on the Amazon*—and now that paragraph in *Nature* seemed to open a portal through which I might pass from my lecture room and laboratory haunts of the past three years right away into the realms of romance.

In due course it was all arranged; I managed to pick up a Vans Dunlop scholarship to ease the strain on the limited means of my father, a retired Indian official.

The necessary arms and equipment were got together: an excellent Express ·45 rifle by D. and J. Fraser, a 12-bore gun—cylinder bore in order to take, on emergency, spherical bullets, a ·476 Webley W.G. revolver and a smaller ·45 ditto. To supplement a large supply of ammunition—mainly No. 9 but also some No. 3 for larger game as well as a few cartridges with spherical 12-bore bullets—reloading apparatus and materials, arsenical soap, botanical drying paper and wire frames, dissecting instruments, microscope and accessories, half-plate photographic camera and plates, clothing suitable for rough tropical wear, field glasses, aneroid and compass, tinplate and soldering apparatus for the construction of tanks to hold spirit specimens.

Eventually all was ready and on Saturday 1st June I set out on my journey.

Everyone is roughly familiar with the continent of South America as it appears on the map: its roughly triangular form, the great range of mountains along its western edge—interpreted by some of us as a great crinkle of the earth's crust being gradually forced upwards by differential movement between the superficial crust and the more rapidly rotating core—and the great estuary on its south-eastern side named by the early Spanish explorers the Rio de la Plata, the River of Silver, suggested by the silver ornaments worn by the aboriginal inhabitants, nowadays vulgarised into the River Plate. To the west of this inlet and extending far to the north and south lies the great estuarine plain known in its southern part as the Pampa and in its northern as the Gran Chaco—both of them names of uncertain origin.

The Pampa-Chaco plain—almost dead level apart from occasional slight undulations—slopes gently to the south-east as is shown by the generally N.W.–S.E. course of its tortuous rivers. These streams flowing from their source in the Andes become greatly diminished in volume by evaporation during their plain course. Eventually they are brought up against the escarpment formed by a great geological fault running almost due north and south along the western edge of Brazil, Paraguay and the Argentine province of Corrientes, and are deflected southwards in the form of a single great river which broadens out into the River

Plate, and is known in its northern portion as the Rio Paraguay and in its southern as the Paraná. Pampa and Chaco have no sharp line of demarcation but it may be said that the southern limit of the Chaco corresponds roughly with the entry of the River Paraguay into the Paraná.

On Sunday, 2nd June, I sailed from Southampton on the Lamport and Holt liner *Maskelyne*—flying the Belgian flag as she was carrying the mails between that country and South America. I found myself in luck in having as a fellow-passenger the Hon. Francis Pakenham, H.M. Minister in Buenos Aires to whom I had potent Foreign Office recommendations. From His Excellency I learned much of interest about Argentina and the Argentines and later on I was to owe him a deeper debt of gratitude.

After a voyage of twenty-five days uneventful except for a call at Madeira—my first glimpse of a really foreign country—and a somewhat narrow escape from foundering in a violent pampero off Cape Frio in Brazil—we reached the River Plate and anchored off Monte Video on 27th June. Near us at anchor was a ship unloading a cargo of Welsh steam coal and it was interesting to think not merely that the purchase money, one pound per ton, would find its way back into Welsh industry, but still more to realise that once the unloading was completed there would be ready available an empty ship to carry back food or raw material to England at greatly reduced freight. The incident served to drive home an important lesson in practical economics as regards the back and forth movements of our overseas trade.

After a short stay at Monte Video we passed on up the wide estuary with its muddy fresh water dotted with green floating islands of camelote (*Pontederia* or *Eichornea azurea*) from the tropical waters of the great rivers away to the northward, and in the morning of 29th June reached Buenos Aires. In these days ships anchored in the Outer Roads, 15 miles out from the swampy shore. Landing was a complicated business—by steam tender, boat and finally bullock cart—the wonderful quays and docks of later days being still non-existent.

For the forthcoming expedition there had been constructed by Messrs Bow, McLachlan and Co. of Paisley, a flotilla of four vessels ('Escuadrilla Pilcomayo y Bermejo') and these, shipped in the form of separate plates, were now being riveted together in a Buenos Aires shipyard. Their construction was as yet not nearly complete, and it would clearly be near the end of the year before they would be ready for our voyage upstream, so I determined to devote the intermediate period to improving my equipment in the matter of practical knowledge of South American natural history. In Buenos Aires itself was the excellent national museum of natural history, where I found a kindly teacher Dr Hermann Burmeister, formerly Professor of Zoology in the University of Halle and author of the well-known *Reise durch die La Plata-Staaten* (Halle, 1861). Then there was the southern bank of the estuary—later to be occupied by the docks and quays of Buenos Aires but at this time a stretch of swampy ground harbouring a wonderful richness of bird life in which I was allowed to wander without interference and make my first acquaintance with the birds of Argentina. Presently there came my great opportunity for the further development of this acquaintance when, through the good offices of the Page family, I received an invitation to visit a friend of theirs, Don Benjamin Carbonell, who owned the estancia of Mate Grande, some 50 leagues to the westward, near the little town of Nueve de Julio, the then terminus of the Western Railway.

CHAPTER II

THE PAMPA

At 7.15 on the morning of 13th August I set out by train from the station Once de Setiembre, so called, as is customary with place names in South America, after the date of an important national anniversary. The first half of our six-hour journey lay through the inner camps as they are called, almost as flat as a cricket field, covered with close green turf, the native pampa grasses having succumbed to the introduced species, and dotted here and there by a solitary Ombú tree (*Phytolacca dioica*) the one tree indigenous to and characteristic of the Pampa. As the Argentine poet Dominguez has it:

Cada comarca en la tierra
Tiene un rasgo prominente,
El Brasil su sol ardiente,
Minas de plata el Perú,
Montevideo su Cerro,
Buenos Aires—Patria hermosa—
Tiene su Pampa grandiosa;
La Pampa tiene el Ombú.

(TRANSLATION)

Every region in the world
Has its special feature,
Brazil its burning sun,
Peru its mines of silver,
Monte Video its hill,
Buenos Aires—beautiful country—
Has its stately Pampa;
The Pampa has its Ombú.

Farther out the country became more undulating, varied by rushgrown lagoons from which rose up flocks of waterfowl disturbed by the train. The Ombú disappeared and the dry ground was now covered by the great native grasses of the Pampa.

Between 1 and 2 o'clock we reached Nueve de Julio, laid out on the usual South American plan with streets running at right angles to one another and surrounding a central square or plaza. The inhabitants with their soft felt hats, poncho and wide trousers or bombachos, went about armed either with revolver or a large knife (facon) stuck in their waist-belt behind—and one could not help correlating with the universal carrying of arms ready for use the admirably courteous manners of the inhabitants.

After waiting a couple of nights in the vain expectation of my baggage turning up, I set out on the 8-league journey to Mate Grande. The vehicle was a four-wheeled trap or 'volante' drawn by five horses urged on by the native driver with wild whoops into a sharp canter or gallop as soon as we got clear of the town.

It was a perfect spring morning, the air crisp and bracing and the ground lightly dusted with hoar frost during the early hours. The road was a mere track across the open Pampa, still uninterrupted by fencing; in parts across smooth green turf where one noticed the curious softness of the motion due to the entire absence of stones, though every now and then one was thrown into the air as a wheel sank into a burrow. Our journey varied in character—a free gallop across an area of level turf, a slow progress through swamp, or a cautious threading our way through a laguna with the water up to the axle-trees. The season being early spring the tall grasses of the Pampa were brown and withered but here and there, most usually on the slope overlooking a lagoon, would be a stretch of beautifully green close-cropped turf with a cluster of burrows like those of the rabbit but twice as big which one recognised as a Biscacha village. It being daytime the rightful owners were not visible but here and there by the opening of a burrow were a pair of charming sentinels, little Prairie Owls (*Speotyto cunicularia*), sitting bolt upright, close together, motionless as statues, except their heads slowly rotating to keep an eye on us as we passed. By the edge of a laguna a flock of dark-coloured Ibises (*Plegadis guarauna*) probed the mud with their long bills, while by another a group of tall Flamingoes (*Phoenicopterus ignipalliatus*) of a rosy pink colour seemed to be dreaming away their existence motionless upon one leg, their long necks coiled up on their shoulders. A great Stork (*Euxenura maguari*) daintily picked its way among the tufts of long grass on the look-out for snakes or other such animals tempted out of their winter sleep by the warm sunshine. Round us flew a cloud of Lapwings (*Vanellus cayennensis*), much like our own though rather larger and armed on each wing with a formidable-looking pink spur, uttering harsh cries of *teru-teru*. Of the smaller birds the most conspicuous were flocks of the starling-like Icteridae, one with a yellow breast (*Pseudoleistes virescens*), another (*Leistes superciliaris*) with breast and gorget of the most vivid scarlet.

Soon after midday that wonderful drive was brought to an end by our arrival at Mate Grande—the estancia house a low-roofed three-roomed cottage, with kitchen, office, and staff accommodation in detached ranchos constructed of poplar trunks filled in with mud and straw, mud floor, and roof thatched with large rushes. The whole was surrounded by a thick plantation or monte, of poplars, weeping willows, peaches, acacias and a laburnum-like tree with drooping racemes of large white flowers.

A cordial welcome from my host Don Benjamin Carbonell—scion of a family of wine merchants well known in Thackeray's day—followed by breakfast of the usual Spanish American kind—soup made with fat freshly killed beef; puchero—the solid content of the soup—beef, pumpkin and rice; asado—roast beef: the whole washed down with vino Carlon—a coarse red Spanish wine. I ought to mention the unfailing aperitif—an admirable cocktail of caña or rum. Don Benjamin was known far and wide as a real artist in cocktails; he was most particular about their trimmings—in particular running a bit of fresh lime (fruit) round the edge of the glass and then inverting it in pounded sugar and taking care that the bottom of the glass contained a cube of pine-apple of the proper stage of ripeness. The eye-opener which he unfailingly brought me before I got up, much as to-day one would be brought one's morning tea, would sometimes be followed by quite a series fitted on to the calls of thirsty neighbours.

My first breakfast at Mate Grande marked the commencement of three months of extraordinary kindness and hospitality accorded to me by Don Benjamin and his companion Don Patricio Kavanagh. My daily routine was pretty regular—the early hours devoted to natural history; breakfast—the beef being killed the same morning—it is well known to those with experience that the most perfect beef in the matter of tenderness and flavour is that which is cooked before rigor mortis has set in; a short siesta, then more natural history, dinner, cards—vingt-et-un, euchre or écarté—bed.

An occasional break would come from a visit to a neighbouring estancia to breakfast or dine and sleep. These involved a few miles' ride across the Pampa—not at the tiresome artificial trot as at home but at the delightfully untiring 'galope' or canter. Happy memories I have of these rides and of how my favourite mount, apparently at the extreme of exhaustion on the outward journey, used to fly like the wind when his head was turned homewards.

In those days the surrounding Pampa was unfenced and the cattle able to wander hither and thither. To keep an eye on them were outposts from the main estancia settlement, each a small rancho inhabited by a puestero. These always received one hospitably but before dismounting it was customary to call out a greeting 'Ave Maria'. In response the puestero would appear, and reply with 'Sin pecado concebido' and one would then dismount, leave one's horse with the reins over its head, and accept the proffered 'mate a la bombilla', i.e. 'Paraguayan tea'—an infusion of the leaves and twigs of the 'yerba mate' (*Ilex paraguayensis*) served in a gourd (mate) and imbibed through a silver tube (bombilla).

Each estanciero's stock bore his particular brand and periodically there took place a rodeo or round-up for the branding of the young animals and the sifting out of aliens belonging to neighbouring estancias. To these rodeos came gauchos from far around and one saw beautiful work with the lasso and the bolas.

After the main work of the day there would be races and an occasional quarrel would be marked by pretty fighting with the facon—a large knife—carried by the gaucho at the back of his waist. The poncho was wrapped round the left arm which served as a guard and also to distract the adversary's attention by flicking the fringe up into his face.

One also heard wonderful tales of the exploits of particular famous gauchos.

Apart from such occasional breaks my time was given to making acquaintance with the plant and animal life of the Pampa, a low-lying undulating land covered with coarse grass and varied in the distance by an occasional monte marking the position of an estancia house. After a long succession of wet seasons the hollows were occupied by shallow lagunas averaging a couple of feet or so in depth. Covering their surface in places was a continuous floating carpet of carmine-brown *Azolla*, here and there intermingled with Duckweed (*Lemna*) with its elliptical green 'leaves'. In the deeper, more permanent, parts of the lagunas were dense growths of a tall rush reaching 9 or 10 feet in height, and much of my time was spent in wading about in these rush-beds and observing the ways of their bird inhabitants. Commonest was the little Rush Spinetail (*Phloeocryptes melanops*) which might be seen hopping nimbly from rush to rush, reaching down every now and then to pick up an insect from the floating carpet of Azolla or Lemna. It first attracted attention by its peculiar voice—several sharp taps as of a slate tapped with a slate-pencil, followed by a long drawn-out squeaking sound like that sometimes made when a tightly fitting cork is twisted in the neck of a bottle. Its nest, firmly tied to a group of rushes about 3 feet above the surface of the water and with blue eggs like those of a hedge-sparrow, is a fascinating little structure built of grass leaves interwoven with extraordinary firmness and covered by a domed roof. Near the top of one side is the little round entrance covered by a projecting eave and leading into a beautifully warm little circular chamber well lined with wool and feathers. Another charming little nest is to be found among the rushes, a cup formed of short bits of grass, etc., cemented together and fixed to a single rush stem about 4 feet above the water. This nest is tenanted by a most beautiful little bird (*Cyanotis azarae*) belonging to the characteristic South American family Tyrannidae; about the size of a wren, its plumage gleaming with the deepest shades of yellow and black and green.

One of the most fascinating inhabitants of the rush-bed was a miniature heron (*Ardetta involucris*) which when flushed would fly away a short distance with characteristically weak flight, its long yellow legs dangling downwards, and would then suddenly vanish from view. On stealthily approaching the spot where he disappeared a slight movement of a rush would sometimes disclose him clinging with doubled up legs to the rush stem, displaying a most beautiful example of what nowadays is called camouflage. The back which the bird presents towards his pursuer is of pale buff colour, traversed by longitudinal dark stripes, the whole pattern harmonising perfectly with the light and shade of the rush-bed. The effect is aided by the whole of the bird except the legs being held in a rigid vertical position, the yellow bill being pointed upwards. The obliterative effect is still further increased by the long loose plumage of the head and neck which conceals their difference in diameter from the trunk on which the feathers are compact and closely fitting. The claw of the middle toe carries on its inner side a beautiful comb which is no doubt of use in dressing the loose fluffy plumage of the neck.

The nest of this charming bird is an inverted cone of radiating bits of rush stem built among the rushes about 6 inches above the water. The eggs are pale green, elliptical and about 1 inch in length.

The four other species of heron met with at Mate Grande included the white and snowy egrets, but these will be dealt with in a later chapter.

Amongst the rushes too were innumerable nests of coots, ducks, grebes and an occasional stork.

Ducks were extraordinarily numerous, especially during the earlier part of my stay at Mate Grande, every bit of open water swarming with them. There was great variety too, eleven different species being identified. Perhaps the commonest was a beautiful little teal (*Querquedula versicolor*) with sober vestments of clear grey, its bill blue grey with a yellow patch on each side. Shovellers, pintails and Chiloe wigeons were also abundant on every laguna. Not uncommon was the Rosybill Duck (*Metopiana peposaca*) a large duck nearly black in colour above and with a large tumid bill of a pinkish red colour. Sailing majestically about some of the deeper lagunas might be seen a pair of Blacknecked Swans or the smaller White Swan (*Coscoroba candida*). Grebes were abundant and on Mate Grande alone I found four out of the five species already known to occur in the Argentine Republic. The finest of these is the Bright-Cheeked Grebe (*Podiceps calipareus*) which Charles Darwin observed at Bahia Blanca. Above dark grey, below snowy white and smooth as the finest satin: each ear covert composed of hair-like feathers of shining metallic golden bronze. The commonest of the grebes (*P. rollandi*) which swarms on every laguna is a much smaller bird with conspicuous white cheeks and dark chestnut underneath.

Mention of the grebes naturally suggests a return to the group of ducks for one of these, the Bluebill Duck (*Erismatura ferruginea*), has adopted the diving habits of grebes and in association with this has assumed remarkable resemblances to them in structure—the wings being reduced in size, the legs shifted back to the hind end of the body, and the breast plumage as satin-like as that of a grebe. The tail feathers, flat and stiff, are without the tail coverts which in other ducks cause a gradual tapering from body to tail. During rapid swimming the tail is spread out in one plane under the surface of the water, ready to be depressed so as to help the act of diving in which this duck is as expert as a grebe, getting into safety below the surface of the water before it is reached by the pellets from a shotgun. When unhurried the duck has a habit of partially submerging itself by compressing its air-sacs and in those circumstances the tail feathers are folded together and bent vertically upwards so that only the neck and head at the one end, and the tail at the other, are visible projecting above the surface of the water.

It was entertaining to watch the process of courtship in this species. The first tentative phase is like that of other ducks, the drake slowly moving his head up and down in the endeavour to get a responsive bow from the female. If she fails to respond he adopts stronger measures; stretching out his neck on the surface of the water he ruffles up his neck feathers and blows up his crop until his neck seems to disappear altogether, and then, folding his remarkable tail, fully spread like a fan, forward over his back, he advances in purposeful fashion towards the female giving his neck convulsive jerks all the while.

It was of great interest to notice that quite another species of duck (*Heteronetta melanocephala*) showed, though to a less extent, the same tendency to diverge both in structure and habits in a grebeward direction. This duck too dives as expertly as a grebe and the male may be seen on occasion swimming with great rapidity, his neck prostrate on the water in front of him and his head almost in the water.

The dry grasslands too presented a wonderful variety of bird life. Carrion feeders abounded as one might expect from the carcasses of stock animals littering the country in all directions—the Carancho (*Polyborus tharus*)—a large almost vulture-like carrion hawk—the most conspicuous. A Pipit (*Anthus correndera*) and a *Wren* (*Cistothorus platensis*) are common in the open camp. More striking were birds belonging to the characteristic New World family of Icteridae: the Yellow-breast and Red-breast already mentioned; *Agelaeus thilius* a smaller bird with a bright yellow patch on each shoulder; and the Cow-bird (*Molothrus bonariensis*), black and rather larger than a sparrow, which attracts notice by its habit of perching on the backs of cattle and sheep and picking off what parasites it can find. That other characteristic New World family of birds—the Tyrannidae—had many representatives at Mate Grande. Perched on a hemlock by the water's edge might be seen a Silver-bill (*Lichenops perspicillatus*) with plumage of pure black except the wings which are equally pure white. The Scissor-tail (*Milvulus tyrannus*) was again common, resembling in general appearance a large swallow but attracting attention during flight by the opening and closing of the prolonged outer tail feathers like the blades of a pair of scissors. The Bien-te-veo (*Pitangus bolivianus*) rather larger than a thrush, with conspicuous yellow crest and belly and a large dagger-like beak. Its loud and harsh voice reiterating the cry *bien te veo* (I see you well) was perhaps the most conspicuous of the bird sounds round the estancia house. The New World Dendrocolaptidae with their dull brown colouring contribute a third characteristic factor to the avifauna of the Pampa. Of them the most interesting is the Oven-bird (*Furnarius rufus*) whose spheroidal mud-built nests about a foot in diameter with a side opening leading to the interior by a cunningly curved passage are often to be seen attached to trees or posts.

The two local game birds, known as partridges, really belong to a quite distinct New World family the Tinamidae or Tinamus. Of these the commoner is the smaller Spotted Tinamu (*Nothura maculosa*). The larger, the Martineta (*Rhynchotus rufescens*) seems less abundant than it really is from its habit of taking to flight only when it is nearly trodden upon. When flushed he usually takes two or three short flights interrupted by short intervals in which he planes along with wings widely spread. In the morning the plaintive whistle—the sweetest bird voice of the Pampa—is to be heard in all directions. It usually consists of five distinct notes, the first two slurred together. Both the pitch and the tempo show considerable variation. Sometimes additional slurs are introduced and sometimes the song is broken off after the second or third note.

The eggs of the tinamus are characterised by their highly polished shells, purplish in the smaller species, green in the larger, and sometimes these birds have the habit, so well known in the European cuckoo, of depositing their eggs in the nests of other species, in this case frequently teal.

The native method of killing a tinamu is to ride round it in a spiral, approaching gradually nearer and nearer until the bird can be struck down by the heavy riding whip.

The lack of trees on the Pampa has been one of the great factors dominating its bird-life but the plantations round the estancia houses have provided a series of stepping stones as it were for the immigration of forest birds. One of the most interesting of these immigrants in the Mate Grande monte was the beautiful Cardinal Finch (*Paroaria cucullata*), of which a pair had nested and produced a family for several successive seasons. The vivid scarlet headpiece had however proved a quite irresistible attraction to the numerous half-wild cats, and the couple of parent birds alone had managed to survive.

Of the Pampa mammals the one that normally first attracts attention is the Biscacha (*Lagostomus trichodactylus*) (Plate I (*a*)) the large rodent already mentioned, about three times the size of a rabbit. The stiff, rigid tail about 6 inches in length which when the animal is running is carried projecting right back, and the hind-quarters, on a much larger scale and higher in the air than the fore-quarters, give the animal a quaint resemblance to a wheel-barrow trundling along of its own accord. The biscacheras or biscacha villages, usually located by the side of a laguna, are recognisable from a distance by the bright green of their grass—a peculiar harsh type called by the natives 'paja brava'. Each burrow has two or three intercommunicating chambers in which the owners remain for the most part during the day, coming out towards sundown. After sunset they may be seen sitting near the mouth of the burrow nibbling the grass. They are extremely bold and inquisitive and have a fair allowance of confidence for, if one approaches slowly and quietly, they will sit at the entrance of the burrow until one gets within a couple of yards. If one stands still or sits down within say ten feet of a biscacha he will sit up on his hind legs and stare at one, giving expression to his indignation and impatience by an extraordinary series of sounds something like the fuffing and groaning of an enraged cat. Finally, if the intruder shows no sign of going away, but remains still and motionless, the biscacha will settle down comfortably and sit still, keeping an eye all the while on the stranger and ready to vanish like lightning into his burrow at any sudden movement.

The biscacheras have other inhabitants than the biscachas. Quite usually by day a pair of the little prairie owls already mentioned may be seen sitting bolt upright as if on guard at the entrance to a burrow. Over the main opening one may see a small opening made by a Miner bird (*Geositta cunicularia*) and this may in turn be taken possession of by a Swallow (*Atticora cyanoleuca*).

When deserted by its regular inhabitants the Biscachera forms a convenient refuge for a variety of other animals, especially the local carnivores—the beautiful silver grey fox (*Canis azarae*); two species of wild cat, the gato montes (*Felis geoffroyi*) something like a miniature leopard, and the smaller *Felis pajeros*; and the Hurón (*Galictis vittata*) a large weasel, greyish yellowish above and dark underneath and next after the skunk the commonest local carnivore. The hurón has a similar defensive arrangement to the skunk and the odour seemed to me still more offensive.

Of all the local carnivores the Skunk (*Mephitis patagonica*) about 23 inches in length, with its silky dark brown coat with a white longitudinal line along each side of the body, and a bushy white tail, is the most conspicuous. He is a beautiful little animal and about sundown may be seen taking his leisurely evening stroll, his white tail held aloft as a signal to all to keep their distance.

A big collie-like dog 'Punch' belonging to a neighbour was fond of accompanying me in my walks abroad and he was one of the few local dogs that would tackle a skunk—gripping it by the back, crunching through its vertebrae with eyes firmly shut and never leaving go until the skunk was dead. Then he would dash into the nearest laguna and wallow in it to get rid of the unpleasant smell, but for the rest of the day he would go about looking very miserable and foaming at the mouth.

During my stay at Mate Grande a single puma, lion as it is called in South America, was killed on a neighbouring estancia. As is well known the puma wanders long distances and is most destructive to stock.

The large Water-Rat (*Myopotamus coypus*), commonly miscalled Nutria (Spanish for Otter) and providing the warm fur known in commerce as nutria, was fairly common. It and the biscacha I used to shoot with the revolver to improve my practice with this weapon in preparation for the Chaco.

Of other rodents apart from species of rats and mice the commonest was a brownish green guinea-pig called locally Cunejo (*Cavia leucopyga*) which forms runs among the grass. Frequenting these and seldom venturing into the open its sense of sight seems to be degenerating and here, as with its relatives in the Chaco, I found a large proportion suffering from cataract.

Of other mammals there still exist in the Pampa survivors of two ancient groups the Edentates and the Marsupials. In former days, long before the advent of man, the Edentates furnished the most characteristic inhabitants of the region, enormous Ground Sloths and Glyptodons. Some years after my stay at Mate Grande I had the thrill of finding and disinterring from the banks of the Rio Quequen farther to the south skeletons of these extinct monsters—represented to-day by the comparatively tiny armadillos. Of the two fairly common species of these the most abundant was the Peludo, or Hairy Armadillo (*Dasypus villosus*), every dry piece of ground being riddled with their burrows so as to make riding at high speed a risky business. I was taught by my friends to ride with a loose seat so as to be thrown clear when the horse came down.

One morning we found that a hen's nest had been robbed during the night and looking around for traces of the marauder we found a newly formed burrow which Punch indicated to be inhabited, so getting a spade we followed the burrow for about 15 feet when there was exposed the scaly tail of a peludo burrowing with quite extraordinary rapidity, almost as if swimming through the soil.

The Mulita (*Praopus hybridus*) is less abundant than the peludo but is yet quite common and may be seen during the day feeding. Both of these armadillos feed on the flesh and maggots of decaying carcasses but in spite of this their flesh, especially that of the mulita, provides a much appreciated dish in the Argentine bill of fare.

A third species of armadillo—the Mataco (*Tolypeutes tricinctus*) also occurs on the Pampa. It is not such a burrower as the other two though it does dig up ants' nests. It is also more diurnal in its habits and runs with great speed on the tips of its toes. Its most fascinating habit is that of curling up into a perfect sphere—the fore and hind parts of its protective carapace being so shaped as to fit perfectly together, leaving only two gaps into which fix exactly the head and the tail each with its armour coating. On the threat of danger the mataco at once assumes the spherical form. If a dog tries to seize it the jaws fail to grip the smooth surface of the animal ball which slips away from their grip and rolls out of the way.

On my later expedition already alluded to I got a single specimen of the Pichy Ciego (*Chlamydophorus truncatus*) (Plate I (*b*)), a charming little armadillo of about 5 inches in length, in the sand dunes near Bahia Blanca. It has very small eyes—whence the Spanish adjective 'ciego' or blind—and the posterior end of its body is covered by a great flattened bony shield, no doubt a valuable protection against an enemy pursuing it as it burrows. Another characteristic feature is its coating of long white silky hair. The special interest, however, of the Bahia Blanca pichy ciego was that this locality was far removed from its recognised home in the neighbourhood of Mendoza—some hundred miles to the north-westward.

Of the Marsupials there were two species of opossum, but I came across only one (*Didelphys azarae*) Comadreja as it is called by the natives, not uncommon in the montes, whence it sallies out during the hours of darkness, doing at times much damage to poultry, not merely by robbing the nests but also by indiscriminate slaughter of the fowls themselves.

Of vertebrates other than birds and mammals I need only mention a couple of reptiles. A large lizard (*Tupinambis teguexin*) called Iguana is common. A particularly large male which I killed measured 52·5 inches in length and its stomach contained ducks' eggs. The other reptile was a large poisonous snake (*Bothrops alternatus*) known as Vivora de la Cruz, from the cross-shaped marks upon its head. I mention it for it was on this creature at Mate Grande that I first made the observation, often repeated in the Chaco, that when coiled up preparatory to striking, it vibrates its tail rapidly against the grass stems, producing a warning whirring sound and clearly indicating the first stage of the evolutionary progress which has led up to the Rattlesnake (*Crotalus*) in which not only does the hind end vibrate, but successive instalments of shed epidermis remain loosely attached to form an actual rattle.

In the monte were many conspicuous ants' nests, each in the form of a low earthy mound 2 or 3 feet in diameter, thatched with bits of leaves and sticks, and inhabited by a large black ant about half an inch in length. Radiating out from the nest were about half a dozen great roads kept smooth and free from obstruction, and along these passed and repassed crowds of busy toilers, irrespective of the conditions of daylight or darkness. The ants going homewards towards the nest carried heavy loads: bits of stick or more often large segments neatly cut out of leaves and often five or six times as big as the bearer. On a warm sunny day there was a constant stream of ants along each of the roads to and from the nest, hurrying along with a wonderful appearance of bustle and energy. These were, of course, the remarkable gardener ants (*Attidae*) which cultivate underground mushroom beds.

The pieces of leaf when carried down into the recesses of the nest are cut into small fragments and used to form the mushroom beds. These are inoculated with a special fungus, the fine threads of which spread through the masses of leaf fragments. The beds are kept in order by special gardener individuals, and these by some unknown method are able to influence the fungal threads so that they produce abnormal growths—little round masses, loaves as it were, upon which the members of the community subsist.

Of other insects conspicuous in the monte were bugs (Hemiptera) characterised by the different colour of their upper and under surface. In one lot the upper (dorsal) surface was leaf green, the lower (ventral) pale brown, while in another lot the same colours were reversed in position the ventral surface being green. In this latter case the insect when over a green background of grass would hang on to the lower sides of a grass stem so as to have its ventral side upwards, and so retain the obliterative effect of the green colouring.

The surface of the Pampa is entirely of fine grained soil, underlaid by a firm rock-like 'tosca', but devoid of the smallest pebbles or stones. This creates rather a quandary for domestic fowls which normally swallow pebbles to serve as millstones to grind up the food in the gizzard. One day a neighbour, Don Luis, came to consult me about a mysterious epidemic which had suddenly caused the death of about thirty chickens. Post-mortem examination provided the solution of the mystery, for each gizzard contained a number of dark grains of carbon. A 'Rapide' charcoal filter had just had its charcoal renewed, the old filling being thrown out. The hapless chickens had swallowed with avidity the hard grains. Each grain after long use had become charged with concentrated poison—condensed poisonous gas—and this set free by the warmth of the gizzard caused speedy death.

Geologists had expressed doubts regarding the origin of the pampean soil, some interpreting it as having been deposited originally from water while others regarded it as a sub-aerial formation. Any doubts as to the former explanation being the correct one were, in my mind, completely set at rest when later on I found the carapaces of the extinct Glyptodons embedded in the banks of the River Quequen invariably upside down—the position that a carcass floating in water would naturally assume, with the heavy carapace downwards.

My stay at Mate Grande extending as it did through the spring season witnessed great changes in the vegetation. At the beginning of October the face of the Pampa was parched and yellow. The grass had not yet begun to sprout; there were no wild flowers except a little yellow oxalis. During October however there came heavy rains and the whole scene changed. Round the estancia houses enormous growths of thistles sprang up, forming within a fortnight dense thickets 7 or 8 feet high. Out in the camp similar thickets of thistles of two different species developed, while a profusion of wild flowers appeared—two species of *Verbena*, one violet coloured the other scarlet, and three species of *Oxalis*, a yellow flowered one in great abundance almost vying with our fields of buttercups at home, another pink flowered and still another white; these, with four different species of vetch, provided an unexpectedly beautiful show of wild flowers.

Another thing of interest to the botanist was seen in the little *Azolla magellanica* which formed a brownish crimson floating carpet over much of the open water. Towards the end of October with shrinkage of the waters the azollas became stranded round the water's edge and shooting their roots down into the soil attained a much more luxuriant growth and now produced an abundance of round sporocarps, at first green but becoming red as they ripened.

On 14th November my stay at Mate Grande came to an end, leaving behind it cherished memories of my first lessons in South American Natural History, of the charming hospitality of the English estancieros and, above all, of the endless kindness of my host, Don Benjamin.

CHAPTER III

THE RIVER PARANÁ

After my return from Mate Grande there ensued a tedious wait of six weeks in Buenos Aires while the construction of the four little ships of the Bermejo-Pilcomayo flotilla was being completed. Of the four it was the *Bolivia* the leading ship of the expedition which was to be my home during its progress.

She was a stern wheel steamer 86 feet in length, flat-bottomed and drawing only 18 inches—a feature which made steering a delicate operation—a shade too much helm making her spin round out of all control. The space of 4 feet or so between deck and bottom was divided into separate holds each with its own hatch, and one of these was allocated to me to use as store and laboratory. The living accommodation was in deck houses—a main saloon aft, and a smaller forward for the crew—the space between being occupied by engine room and stokehold. On the upper deck was the commander's cabin, and immediately in front of it the wheelhouse with speaking tube down to the engine room. Above the wheelhouse was mounted the ·45 Maxim gun, which must have been one of the very first automatic machine guns to be taken on active service.

Two other vessels, the *General Paz* and the *Caa guazú*, were similar to the *Bolivia* in their general construction but larger—200 feet in length.

The fourth member of the flotilla the *Perseverancia*, or *Heave-up* as she was usually called, was of quite different construction. Very strongly built with a single screw-propeller she was fitted with shears in the bow for pulling up snags, while scattered about her deck were various other pieces of apparatus—including a powerful centrifugal pump for washing away sandbanks and a circular saw; she was in fact a floating workshop.

Towards the end of December the ships were nearing completion and I shifted my quarters from the Hôtel de Provence to the *Bolivia* which was now afloat in La Boca del Riachuelo—the only harbour of Buenos Aires in these days. I lived alone on the boat and took my meals ashore at an eating house kept by a Swedish woman. The dockland of La Boca was notorious for the queer people who inhabited it, many of them Neapolitans, many of them persons who had suffered 'una disgracia'—a misfortune—the common euphemism for having human blood upon their hands. However, I found them to be, like the modern American gangsters, well mannered and gentle and by no means uninteresting. One evening when they crowded on to the *Bolivia* for a farewell celebration has always remained a vivid memory.

At last all was in readiness, our ship's company arrived and at 4.30 a.m. on 29th December orders were given to have steam by 7.30. It was midday however, before we eventually emerged from La Boca and steamed away to the north-west towards the delta of the Paraná. For a couple of days we lay at anchor off San Fernando in a small tributary of the Lujan, and finally about 1 p.m. on New Year's Day a start was made.

Wednesday, 1st January: Ropes were cast off, the engines began to move, but the *Bolivia* remained stationary. The water had suddenly gone down, as the waters of the Plate are apt to do with change of wind or of barometric pressure, and we remained stuck fast on a mudbank. However, we jumped overboard and the *Bolivia* was soon pushed off and afloat. It was about 3 o'clock when at last we steamed off and after a short run out to seaward in the vain hope of sighting the rest of the flotilla we turned back, and about four o'clock entered the 'Arroyo del Capitan'—one of the minor channels of the Paraná delta and which, being sheltered by the islands, was to be our route towards the main stream. The Capitan we found to be a beautiful winding creek from 30–60 yards in width and with about a couple of fathoms of water. The banks, but slightly raised above the water's surface, were clothed with a dense growth of vegetation, luxuriant though with little variety—Willows and Poplars being the most abundant trees. The scenery in a way was pleasant but soon became monotonous, and its beauty was rather spoiled by the muddiness of the water. To-day being a feast day the river was crowded with steamers, boats, and launches containing picnickers and holiday makers. One launch, packed with people, happened to pass just as we rounded a sharp turn. Our stern swung rapidly outwards and had it been a second sooner would infallibly have crushed the launch to matchwood. Those on board got a bad fright: the men shrieked and were obviously badly scared while the ladies, on the other hand, appeared to remain quite cool and selfpossessed. As evening wore on we emerged from the Arroyo del Capitan upon the Paraná de las Palmas, the south-westernmost of the main channels of the delta, about three-quarters of a mile in width, bounded by heavily wooded shores; its surface was dotted with great floating masses of camelote with its spikes of beautiful blue flowers. We continued our journey up this channel, hugging the bank so as to avoid the force of the current and, eventually, about ten o'clock, drew in to a narrow creek on the Eastern bank and anchored for the night. Soon after anchoring a heavy splash was heard—evidently of some animal, no doubt a Carpincho or Capybara, jumping in to the water. To-night we had our first real experience of Mosquitoes. Close to the bank as we were they swarmed around us in millions: great, black creatures of unparalleled ferocity. Captain Page said he had never seen anything like it. The surface of the water around the *Bolivia* was completely hidden by camelote, enormous quantities of which were being brought down by the river, at present in high flood. Along the banks today a gorgeous Icterid has been conspicuous—resembling a blackbird but with head and neck of the most vivid scarlet (*Amblyrhamphus holosericeus*).

Thursday, 2nd January: At daylight we weighed anchor, and continued on our way up stream: the banks still low and marshy and covered with a vegetation becoming more purely indigenous than that nearer Buenos Aires. The stream was bordered by a fringe of great rushes, while the islands were for the most part covered with woods of Ceibo (*Erythrina crista-galli*), a smallish tree with crooked, much branched and somewhat oak-like trunk, trifoliate leaves, and conspicuous from its gorgeous flowers like gigantic pea flowers of rich and beautifully shaded grades of scarlet. The surface of the islands generally was covered with rank vegetation and there were many beautiful flowers: the deep red Ceibo, yellow Cannas and white Sagittarias. We soon left the main channel of Las Palmas for the Baradero, a narrow stream of much beauty, reminding one not a little of one of the larger midland streams at home. Shortly before entering it the South-West bank changed its character—the low and swampy ground of the delta being replaced by a tall vertical bluff marking the edge of the mainland. The Baradero skirted along near the base of this while the main channel took a wide sweep to the Eastward. We thus steamed along the chord of an arc, and greatly shortened our route thereby. The Baradero was thronged with birds: Storks in great numbers; big Cocoi herons; Night herons; White and

Snowy Egrets; Cormorants; a Grebe or two; many Chajás and Ibises; Caranchos and Chimangos: while the scarlet headed Icterid was abundant and conspicuous. Until near the town of Baradero the banks were well wooded, comparatively dry, raised a few feet above the level of the water, but after passing the town the stream threaded its way through an immense marsh swarming with mosquitoes. Darkness fell while we were in the midst of this slough of despond and it was at times scarcely possible to distinguish the channel ahead of us. The flat expanse of swamp stretched out for miles on either hand; the whole sown with bright points of light from fireflies which scintillated and rose and fell. As we steamed along in the darkness, the bright glare from our furnace attracted myriads of fish and during the evening over a dozen of the large salmon-like Dorado (*Salminus brevidens*) leapt on board, several going right down into the stokehold: and looking over the side we could see multitudes of these fish keeping exact pace with us as if attached to our ship by invisible wires. After about three hours spent in traversing this dismal swamp, there was at last visible in the distance a silvery streak shining in the moonlight which, as we approached nearer, assumed the appearance of a great lake studded with innumerable islets—and this was our first glimpse of the mighty Paraná guazú. We steamed on till we neared the point at which the Baradero again joined the main stream and there dropped anchor for the night.

Friday, 3rd January: Day broke wet and dismal, with a high wind blowing, so Page deemed it advisable to remain within our shelter in the mouth of the Baradero. I employed my time by going ashore to collect. The ground was low-lying and marshy and the vegetation luxuriant. There were several fine flowers but the finest was a Hibiscus with immense pink bell-like flowers, which was abundant. In the afternoon the wind moderated, we cast off our moorings and continued our way up the Paraná. The river was at this point about four-fifths of a mile broad, with a current of about three miles an hour. The high bank retreats from the river at the town of Baradero, but rejoins it at Obligado, as a steep bluff about sixty feet in height; evidently the truncated edge of the Pampa, with grassy pasture on the top and here and there a solitary Ombú. About ten o'clock in the evening we arrived at Rosario, the second city of the Argentine, and were there joined by the *Caa guazú*.

We lay off Rosario for nearly a week during which we took stores on board while I got my laboratory in the hold into order. The ship's company received a new recruit, the son of a General

Virasoro who had been assassinated while Governor of the province of Corrientes. Young Virasoro had been educated in England, later had held a commission in the Argentine army and was now rather down on his luck. He had had a colourful career and made interesting conversation.

Two other prospective members of the ship's company—Kennion and Daniel—had not turned up, so we left without them on 9th January after taking six tons of coal on board.

The day by day progress along a great river like the Paraná shows, apart from navigational details, much monotony and sameness, so I will for the time being drop my diary and give a more general account of our voyage until we reached our Chaco base.

I had formally no official status but naturally had to take my share of work with the other members of the ship's company and as soon as the Commander realised that I was accustomed to boat work he allocated me to pilotage duty. Charts of great rivers like the Paraná are of course useless owing to the constant changes of the channels, and navigation is done entirely by reading the indications which the surface of the muddy water provides. Page was a skilled expert in such pilotage and in the early days of my training he would sit just outside the wheelhouse conning the waters ahead and interpreting to me its indications until I too developed a certain amount of skill in this fascinating art. When we saw ahead of us a wide area where the water appeared uniformly shallow and without any indication of a navigable channel, we would drop the anchor and go ahead in the small boat seeking out a possible channel by sounding with a long cane. In navigating upstream we normally hugged the bank, but when going downstream we kept in mid-channel so as to take advantage of the greater strength of the current.

During the hours of daylight while we were under way I was on practically continuous duty in the wheelhouse. At night when at anchor or moored to the bank we took two-hour watches in turn and the others agreed to my suggestion that we should keep to the same hours so as to get accustomed to turn out at the same time every night. My watch was 2–4 a.m. and remained so throughout the expedition.

Friday, 10th January: Everyone nearly driven mad by the mosquitoes last night. During sleeping hours, inside one's mosquitero, one was comparatively safe, but on emerging to dress we were each assailed by hungry swarms, and all exposed parts were in a moment black with mosquitoes. It was quite out of the question to dress in the saloon, so I got out on deck where things were rather better owing to the breeze. Yesterday we were informed that Friday was to be washing day and that everyone had to wash his own things. Even the Commander said he meant to do his own washing, so I commenced mine: I scrubbed away industriously in sailor fashion with a hard brush at the garments spread out on the deck but didn't prove skilful at the art.

We had been all morning lashed alongside the *Heave-up* which is slower than the *Bolivia.* I had just finished my washing when a sudden shock nearly threw us all overboard: the bollards to which the hawsers of the *Heave-up* were attached carried away, and we shot ahead—leaving her stuck fast on a sandbank. It took us a couple of hours' hard work to get her off; towing was of no avail: eventually a line was sent ashore and made fast to a tree, and by hauling with the powerful steam winch she was at last got afloat. A disaster, however, happened during the operations. We had a steam launch towing behind and as our engines were going astern this was drawn underneath our starboard wheel and a large segment cut clean out of her side. With the derrick of the *Heave-up* the maimed launch was hoisted into a large open boat which was also towing astern, and we proceeded linked together as before. Later in the day we got aground again at a point where the water shoaled very steeply, there being three fathoms on our bow and only 18 inches on that of the *Heave-up.* We managed, however, to get off this time by merely backing astern. About 4 p.m. we passed Diamante where the National Naval School is situated. At this point, which marks the northern limit of the Paraná delta, the left bank of the river changed its character. Hitherto its eastern side had been bounded by low and swampy alluvial islands, but now the mainland of Entre Rios came down to the water's edge—forming an escarpment of a couple of hundred feet in height. This range of cliffs was beautified every here and there by mantling clusters of luxuriant vegetation which is now assuming a semitropical character: beautiful convolvuli and other creepers entwined the trees. Birds were numerous. The little dark green Kingfisher (*Ceryle americana*) with white collar and red breast pretty frequent. Here and there upon a tree might be seen a group of white egrets with a Cocoi heron or two. In the cliffs were many holes inhabited by birds of various species. Amongst other birds Parakeets were flying about, the first we had seen so far. Late in the evening we again went aground in mid channel. It took us some time to get off and as it was now late we went alongside the bank and made fast.

Next day we got under way about 4.30 and about eight o'clock anchored off the city of Paraná. The river is here several miles wide though one never gets a clear view right across owing to the innumerable islands. Over them, however, in the distance one could just see the topmasts of the shipping of the port of Santa Fé on the western bank of the river.

Sunday, 12th Jan.: Today we had the highest temperature so far encountered, 100° Fahr. in the shade. It was by no means unpleasant; and Pool, who acts as steward and photographer, and I had a delightful bath sitting on the lowest floats of the paddle wheels while the almost tepid water rushed over us.

The great heat turned out to be the forerunner of a storm and about five o'clock it began to look very bad: the sky became rapidly covered with a pall of dark and ominous-looking cloud. Then the wind rose to hurricane violence whirling clouds of dust into the air. We paid out more cable; and owing to the suddenness of the storm had all to look alert in rigging up curtains and making things shipshape. The storm continued with the greatest fury for about an hour—incessant flashes of lightning playing all round us, but the *Bolivia* rode it out splendidly.

When ashore at Paraná we heard a sound which later became familiar, a shrill longdrawn whistle, apparently of a railway engine. We heard these all around—no railway engines, however, being visible—and it was some time before we realized that the sound was produced by an insect—a large Cicada about two inches in length, of a brownish green colour.

After a couple of days at Paraná we proceeded northwards, the left or eastern bank of the river retaining for the next hundred miles the same character, a precipitous cliff about 120 feet in height. Birds were abundant: large flocks of duck—particularly Rosybills (*Metopiana peposaca*) and Whistling Ducks (*Dendrocygna fulva*); a small flock of Cassin's Tern; an occasional Spoonbill. A call at the Colonia Hernandá to order stores gave an opportunity for a little botanical collecting.

16th January: In the earlier part of the day we coasted along the Western or Chaco bank; luxuriant semitropical vegetation; many dead trees covered with mantling creepers. Of birds there were many Caranchos and White Egrets, and we also saw our first Jacaré or Alligator. The body of a man strung up to a tree emphasized the distance from one's homeland.

We called in at the sleepy little town of Goya, situated on a narrow by-channel of the Paraná, almost the only sign of life being a gang of unpleasant looking criminals working in chains.

The water here was teeming with small fish, and crowds of Terns flew hither and thither, hovering in the air, swooping down to the water's surface and rising again with a captured fish. It was interesting to see that the fine grained alluvium forming the high eastern bank showed beautiful columnar structure, as perfect as that of the basaltic rocks of Fingal's Cave or the Giant's Causeway. Though this prismatic jointing is theoretically the natural thing to develop as a bed of mud shrinks in drying, it is seldom seen in nature owing to the rarity of mud being absolutely homogeneous.

It was a tiring day as I had long turns at the wheel.

18th January: Maynard, the bos'n, has been rather refractory and generally objectionable of late and he has roused the Commander's ire especially by his habit of expectorating skilfully at flies upon the white woodwork. This morning he had been lying about in one of the holds doing nothing and when reprimanded was inclined to be cheeky. So I was ordered to steer in to the river bank and the next thing I saw was Maynard being bundled ashore and his kitbag thrown after him.

In the afternoon we arrived at Bella Vista, some 54 miles above Goya. Virasoro and I went ashore and walked through the town; not a soul visible in the streets and the only sign of life the doleful sound of a band in the barracks practising what seemed to be a funeral march. However, on the way back we discovered a little café and sitting outside it we enjoyed the beautiful panorama spread beneath—the wide expanse of the Paraná looking like a mighty lake studded with innumerable wood clad islands. The town had fully earned its name of Bella Vista.

Sunday, 19th January: Up till now we had used coal in the furnace but to-day we changed over to the wood which was to be our fuel from now on. Nearly all day we were kept busy loading and stowing 1900 logs which filled both bunkers and piled up the deck as well. And after getting under way about seven o'clock, we discovered that the new fuel had introduced a new complication into our navigation for the hot cinders issuing from the funnel made it necessary to keep continuously sluicing our awning and decks with water to keep them from catching fire.

Next day, after a long day's steaming, threading our way between the islands, we arrived at our Chaco base, Barranqueras, or, as it had come temporarily to be called, Puerto Juarez Celman, after the not-yet-inglorious President of the Republic. Puerto Juarez Celman, which remained our home port for over a month—broken into by a couple of trips down river—and also serving as the port of Resistencia, the military capital of the Chaco, consisted merely of a small group of houses and stores on the West bank of the river nearly opposite the town of Corrientes.

On the day after our arrival we took on board 300 logs—a sufficiency of fuel to take us to a woodcutting establishment or 'obraje' where we should get a further supply. The obraje was about nine leagues distant, on the banks of the Tragadero, a small arroyo or creek flowing into the Paraná, about 50 yards in width and 4 fathoms in depth, winding hither and thither between high banks crowned with gloomy forest. The trees were comparable in size with ordinary large trees at home, mostly with dark and sombre foliage and many bearing numerous epiphytes and lianas. Sheltered by the forest the water had a surface smooth as ice, reflecting as in a mirror the birds flitting over it, numberless kingfishers of varied hue and size, an occasional white Egret, dark and funereal looking cormorants.

22nd Jan.: Loading logs occupied the crew most of the day, while I did some collecting. The morning I concentrated on birds as they are to be found moving about most freely before the heat of the day sets in. Two species were especially conspicuous, the Ani (*Crotophaga ani*) a black cuckoo with a clear piping cry something like that of the Curlew, and the beautiful snowy white Viudita Tyrant or Little Widow (*Taenioptera irupero*). The vegetation was luxuriant—open grasslands varied by islets of forest consisting of large hardwood trees such as the Quebracho—on the trees many Gesneraceous and other climbers, while underneath was an almost impenetrable undergrowth of Cacti, an Opuntia, and in particular a Bromeliaceous plant allied to the pineapple and called Caraguatá-i.

Sunday, 26th Jan.: This morning we crossed the Paraná, here over a mile wide, very deep and with strong conflicting currents which give their name to the city on the eastern bank, San Juan de las siete Corrientes, or, as it is commonly called, Corrientes. Our visit was for the purpose of loading 2000 logs of wood for an emergency dash down stream. The *Caa guazú* had been towing a large hulk full of stores for our base at Barranqueras, but she arrived there without her tow, reporting that the hulk had been left firmly stuck on a sandbank. The Commander of the *Caa guazú* was promptly placed under arrest and the *Heave-up* was despatched down stream to try to pull off the hulk. After taking our supply of fuel on board we followed suit.

Most of the next day was spent steaming down river keeping in midstream so as to take full advantage of the current and in the evening we arrived alongside the hulk, which had floated off quite easily and was now anchored in about 3½ faths.

Tuesday, 28th Jan.: We prepared to start upstream but were much delayed by the huge old-fashioned anchors of the hulk, deeply embedded in the mud and requiring a couple of hours hard work at the hand winch to get them up. Then at last the hulk resumed her progress upstream, this time in tow by the *General Paz*. The *Bolivia* got under way later in the day and it was nearly dinner time when we arrived at Goya, steaming right up the narrow channel on which it stands. Our arrival there excited great interest amongst the townsfolk, to most of whom the sight of a steamer like ours was a new experience. Quite a crowd of the inhabitants collected on the bank, squatting down and watching all our movements. We gave them a real fright by sounding the siren—beginning with low threatening mutterings and savage growls, rising finally to a piercing and earsplitting shriek. In the evening many of the ladies of the place came down to see us, and they quite bore out the reputation of Goya for its beautiful women.

After dinner Daniel went ashore to cater and returned with a supply of biscuits, melons, sweets, and various other delicacies not included in the *Bolivia's* stores.

Just before reaching Goya we again saw an alligator sunning itself by the bank, and fired at it but not to its detriment. Many of the islands in the Paraná here were covered with a dense growth of a tall, somewhat poplar-like Composite tree called palo bobo (*Tessara absinthoides*) growing so close as to be practically impenetrable, and many twined about by Convolvuli with beautiful purple-white flowers and palmate leaves.

29th Jan.: Just as my watch was finishing at 4 a.m. a violent squall came down on us with great rapidity, tearing the *Bolivia* from her moorings and driving her out into the channel. Before going very far, however, we got an anchor out with lots of cable and safely rode out the remainder of the storm. While it continued the wind was very violent and accompanied by heavy rain, but after an hour or so things quietened down and we were able to return to our moorings and start the day's work by replenishing our supplies of fuel.

2nd Feb.: A dull morning, cold, bleak and shivery. I had my usual watch from 2–4 a.m. At 4.30 we got under way and I was at the wheel till 11.20 p.m. making a 23 hour day before I got to bed. In the afternoon Kennion passing the boat round the stern was pulled overboard by the sudden tautening of the painter. Fortunately he retained his hold of the painter and managed to clamber into the boat and we were able to throw a line to him after a little trouble.

3rd Feb.: After about an hour's run we were getting short of fuel so tied up to an island opposite Empedrado to cut wood. It was a fine sunny day and many small brown lizards ran hither and thither while tiny green Humming birds hovered in the sunshine. Of other birds, Caranchos, Chajás, large pigeons, very many Tijeretas and Dendrocolaptids, especially Red Oven birds and *Phacellodomus striaticollis*. The last mentioned is a common bird, always to be seen in pairs, and its song, like that of many of its relatives, is [as described in Chapter VI] a duet sung by male and female, the two birds singing in concert, one always a few notes ahead of the other.

Late in the evening we arrived back at Puerto Juarez and found Daniel awaiting us, looking considerably the worse for wear after a series of adventures in Corrientes.

4th February: Next day we again started downstream, this time bound for La Paz to pick up Mrs Page—the Commander's wife. About nine in the evening while entering a narrow channel, the water shoaled suddenly, and before the engines could be stopped and reversed we were hard aground, with only 20 inches of water at the bow and less amidships. Tried to pole her off but failed; jumped overboard but still failed; so we got an anchor into the small boat and Pool and I managed to take it to deep water though with some difficulty as my oar snapped at a critical moment. We narrowly escaped being carried away downstream by the strong current but fortunately were able to reach the shore and tow the boat upstream to the *Bolivia*; not a pleasant task, clambering in a state of nudity through all kinds of prickling and stinging plants swarming with bichos colorados, the local variety of the Harvest Bug.

Back on the *Bolivia* again and hauling hard on the anchor cable, we at last got her off and steamed away triumphantly down stream. I was at the wheel till 4 a.m. and had only an hour in bed. We steamed on all day, and late in the evening anchored close inshore on the North side of La Paz—a total run of about 277 miles.

After a couple of days at La Paz, we started northwards again, stopping to pick up 2000 logs of ñandubey (*Acacia cavenia*) at the neat little town of Esquina. A fall of ¾ inch of the barometer and a most forbidding appearance in the murky blue clouds all round the horizon seemed to portend something unpleasant, so a few miles from Esquina we turned into a little creek and anchored in the midst of a great expanse of Polygonum and tall grass where we should be well sheltered from any sea and at least partially so from the wind. Apart from appearance, the evening was pleasant and I enjoyed a nice bath before turning in.

11th February: During the night we had a terrific hurricane. Just about midnight I was roused by a tremendous blast striking the ship and threatening to blow us away altogether, so I immediately dashed forward to see that the anchor was all fast. However, stepping on a tarpaulin stretched over a hatchway from which the hatch had been removed by one of the crew sleeping below, I fell right through to the bottom of the ship, bruising myself pretty badly. I had to retire for a bit and meantime the tempest raged with great violence, threatening to tear away the deckhouses altogether and making us thank our stars that we were in such a secure haven. The gale continued, though with gradually decreasing violence, till well on in the morning.

About 8 a.m. we prepared to get under way but found this a matter of some difficulty as we were lying in a dense mass of vegetation—tall Polygonum and coarse grass growing up from a depth of 6 to 9 feet to a height of 2 feet above the surface. After 4 hours' hard work trying to free the ship, we had at last to send out the boat with anchor to haul her off. Quite a sea was still running as we steamed away upstream.

12th February: Coasting along an island about 20 miles below Goya clad with magnificent vegetation, we saw bounding about among the tree tops a troop of the beautiful black Howler monkey (*Mycetes caraya*)—1° farther South than had been given by the great naturalist Azara as the southern limit of this species. Other interesting things seen were the small Kingfisher (*Ceryle americana*); an apparent planarian worm about 3 inches in length which turned out to be in reality an extraordinary slug flattened out so as to look like a planarian; many little brown-green lizards; and myriads of dragonflies.

13th February: The vertical bank of the river between Goya and Bella Vista, about a hundred feet high, shows beautifully in almost diagrammatic form the denuding action of surface water. Below the thin superficial layer of soil the bank is carved out into a series of small pointed buttresses giving an impression like a row of Gothic arches. Below these, the buttresses larger and more obtuse; some of the larger with their sides carved into smaller secondary buttresses.

At Puerto Juarez a stay of ten days was occupied in final preparations for our expedition into the Chaco, stores being taken on board and stowed while I took advantage of the opportunity to add to my collections.

CHAPTER IV

RIVERS PARAGUAY AND BERMEJO

26th February: Everyone has been busily occupied during the last two days in making final preparations for our start, as after leaving Puerto Juarez the expedition will be dependent on its own resources. The military detachment arrived on the scene a few days ago—50 men of the sixth and ninth regiments of cavalry. They are without their mounts though having their saddle gear for use as a bed. The Argentine military saddle or 'recado' composed of a leather-covered wooden body when in use surmounts a thick cushion of rugs, and sheepskins. It is the most admirable possible for expeditionary purposes, the thick pile of underlying skins and rugs effectively preventing the occurrence of sore backs even on the longest journeys, while the concave curvature of the wooden body makes it a first rate pillow. The bed itself consists simply of the rugs and skins spread out on the ground. The troopers are clad in blue uniforms with wide 'bombachos' tucked into top boots and peaked caps 'à la Zouave'. They are armed with Remington carbines, excellent simple weapons but objectionable to the English eye in their conspicuous brightly sandpapered steel barrels; and heavy sabres. In addition to these each trooper carries a large knife or facon in his waistbelt behind. Altogether a sufficiently tough and weather-beaten looking lot; no doubt well qualified for the rough work of Chaco warfare. Colonel Racedo their chief is a tough looking soldier too, with a heavy record of Chaco campaigning. He is immensely proud of his men whom he characterises as regular 'leones' in courage. So far the soldiers have been kept busily engaged in getting stores on board the boat and in slaughtering bullocks to provide beef for the voyage. The bullocks are slaughtered in ordinary Gaucho fashion and the beef converted into charqui, that is to say cut into thin strings or slices from a quarter to half inch in thickness which are hung up upon lassos in the air and sunshine until quite dry and hard.

On board the *Bolivia* itself there have been lots of odd jobs to finish. In my own department I have got 50 gallons of strong alcohol for preserving specimens; while a range of tanks for the same purpose is to be fitted up at Asunción. I have also purchased a big non-descript dog Turco described by its owner as a veritable lion in tackling big game.

Our crew on board the *Bolivia* are rather a scratch lot, Daniel and Virasoro have disappeared, while Munsey the very efficient engineer has decided that he has already had enough of it and has accordingly departed for Buenos Aires. In his place Page has engaged an elderly Scot named Henderson who had sailed with him before but who looks hardly robust enough for the strenuous times ahead. Under him as second engineer is an Italian named César.

The armament of the ship's crews consists of Winchester repeating rifles with sword bayonets. On the *Bolivia* itself which will have to bear the brunt of the fighting, if there is any, we have in addition the new ·45 inch Maxim machine gun mounted on the top of the wheel house, an odd Lee rifle, and Colt Navy revolvers. The supply of ammunition on board seems enormous.

Later in the afternoon we crossed over to Corrientes accompanied by the *General Paz* and the *Heave-up*; took fuel on board, and steamed up stream for a few miles before stopping for the night.

27th February: Early this morning we passed the junction of the rivers Paraguay and Upper Paraná; the latter though much the larger comes in as a side branch, the Paraguay and the Lower Paraná being in line with one another, flowing from North to South along the boundary line between the high ground of Paraguay, Corrientes and Entre Rios on the East, and the low-lying Chaco-Pampa plain which extends westwards to the Andes.

The Upper Paraná at its mouth was quite free from islands, presenting an uninterrupted expanse of water right across, while the Rio Paraguay—looking very narrow to our eyes so long accustomed to the mighty Paraná—was thickly studded with islands. The banks well wooded—in many places clothed with a dense growth of the so-characteristic Palo bobo, and elsewhere, especially where the banks were high and steep, with a luxuriant growth of tall canes with leaves arranged fanwise.

28th February: To-day we remained at Puerto Bermejo—or as the Paraguayans call it, Timbó, or more accurately, Nuevo Timbó, the original towns of that name being on the Eastern or Paraguayan bank—a small village of mud huts about six miles below the mouth of the Rio Bermejo. I had a great time collecting in the luxuriant forest to the North of the village, made the more easy by there being distinct pathways through the forest. Among the birds were two of special interest—whose acquaintance I made for the first time. One was a Surucuá (*Trogon surucua*)—one of these strange Trogons—with, in the case of the male, beautiful metallic green plumage—which seem to dream away the greater part of their existence, sitting about in the branches of the forest trees. Their beautiful plumage reaches its highest development in the well known Quetzal, the national bird of Guatemala.

The other was *Xiphorhynchus lafresnianus* [Plate VIII (*a*)], a member of the tree creeping Dendrocolaptids, differing from all its congeners in the extraordinary length of its slender downwardly curved bill, which forms an admirable probe and forceps by which an evasive insect can be drawn out from a deep recess. When an insect is seized by the tip of the bill the bird will slide the latter down against the edge of a branch so as to push the insect to within reach of the unusually short tongue and enable the act of swallowing to be performed.

At this point it was decided to take the flotilla for a short trip up the turbulent waters of the Rio Bermejo and thereby subject the vessels to a searching test before starting on the unknown waters of the Pilcomayo. In the evening Nelson Page, son of the Commander, who was now in command of the *Heave-up*, came round to invite me to go on with him to the Bermejo instead of waiting for the *Bolivia*. I gladly accepted his invitation and it was decided to start next morning.

1st March: About seven o'clock I carried my rug and collecting apparatus on board the *Heave-up*. The *Perseverancia* is altogether a remarkable looking craft. Rather shorter than the *Bolivia*, she has high bluff bows and is built so as to be of enormous strength. Over her bow is a stout derrick with powerful winch and cable for hauling up snags by main force. Below she has a full equipment of auxiliary apparatus, including diving suit and airpump. She was really designed specially for service on the Bermejo and for the removal of the snags, sandbanks and innumerable other obstacles which were known to present formidable difficulties to its navigation. Unlike the other vessels of the flotilla, the *Perseverancia* is driven by a screw propeller and she falls behind them considerably in point of speed.

About 7.30 we cast off and proceeded on our way Northward. Apart from the narrowness of the Rio Paraguay as compared with the Paraná there came a pleasant change in the scenery, the Eastern or Paraguayan bank being parklike and less densely wooded than the Western. The little village of Humaitá with its red-roofed cottages made a pleasant contrast with the monotonous architecture of Argentina to which we had been so long accustomed.

As we neared the mouth of the Bermejo a surprising change came over the waters of the river, which had been uniformly muddy hitherto throughout the whole length of our voyage, for there now came a marked difference, the muddy water being restricted to the Western part of the channel. Actually the waters of the Paraguay itself are dark and clear; those of the Bermejo on the other hand like liquid mud which pours out into the Paraguay river, tending at first to flow under its less dense waters, but coming up to the surface in curls and blotches in the gradual process of admixture.

About nine o'clock we reached the very unimpressive mouth of the Bermejo—a mere 70–80 yards in width, between low and flat banks clothed in the usual dense growth of Palo bobo. Near the water's edge lay several large Jacarés basking in the sun.

Once within the entrance, the river became wider—100 to 170 yards. The banks still continued low lying and evidently subject to frequent inundation. At this time the Bermejo obviously was in high flood and it was slow work steaming against the strong current which appeared to vary between 4 and 6 miles an hour. The water itself was as mentioned before practically liquid mud: it was however, quite drinkable, with a less sweet taste than that of the Paraguay and one distinctly felt the mud against one's teeth when drinking it. When filtered it showed a faint pinkish colour.

As we proceeded up river the banks increased in height and became crowned with luxuriant forest of Lapacho, Pacara, Espinillo and other trees. Towards sunset a pair of Toucans (*Rhamphastos toco*), bizarre in their dark plumage with great red and yellow bill stretched straight out in front, flew across the river just ahead of us and during the evening several more were seen perched in the trees by the river bank. In the evening I shot a Jacaré lying on a steep bank close by the water and after some trouble got it on board. It measured seven feet four inches in length.

2nd March: Being free from the cares of navigation I was able to give most of the day to skinning and dissecting the alligator. The deep grooves in the skull, filled with clear jelly and covered in by thin skin, recalled the sensory tubes of fishes and suggested an interesting laboratory problem to enquire into their real meaning. Apart from its scientific interest the Jacaré ministered to our comfort during the day as both at breakfast and dinner our main dish consisted of fried steaks from its tail—excellent, like good white fish though with a harder consistency.

The course of the Rio Bermejo is an extraordinarily tortuous one, especially near the mouth as farther up long winding reaches take the place of the sharp and erratic turns lower down. While traversing the latter we were constantly impressed by the immense activity of the river as a geological agent. The outer banks of the curves were steep and crumbling—here active erosion was taking place, while the inner banks were low and sloping sand-banks where deposition was taking place. Often on our voyage up the Bermejo the stillness of the night was broken by a dull roar in the distance which told us of the sudden precipitation into the river of an enormous slice of the river bank. In this way immense masses of many thousands of tons weight with all the superjacent growth of forest trees and other vegetation are swept away by the waters. The trees being mostly of extremely dense wood of high specific gravity sink either completely to the bottom or the roots only become embedded in the sand and mud while the trunks project down stream and form the much dreaded snags. After some practice the eye becomes able to detect the hardly perceptible disturbance of the water caused by a snag within a foot or two of the surface. It is of course in going up stream that snags constitute the greatest danger to navigation as when running with the stream the vessel most usually presses them down so that they pass harmlessly underneath. The soil itself is swept more or less completely away and accounts for the enormous proportion of solid matter held in suspension by the Bermejo water and which give to the river its name (Sp. Bermejo = vermilion). In the longer reaches comparatively little erosion takes place: although the banks are here steep they are covered with a thick growth of grass and cane brakes.

The scenery now with its parklike alternation of grass and woodland remind one much of some of the larger rivers at home. The general effect has very little wildness and it is quite difficult to realise that we are in a land inhabited only by wild beasts and wilder Indians. It appears admirably suited for colonization, is well watered and has a delightful climate the only drawback seems to consist in the superabundance of insect pests. From sunset until well on in the morning the air is filled with mosquitoes of many species each fiercer and more relentless than the other. A few hours after sunrise the mosquitoes in great part disappear but those that remain are joined by several allies, incomparably most detestable of which is a horrid fly called by the natives Viuda or widow, let us hope merely on account of the coloration of the wings giving it the appearance of wearing a long black cloak. The bite of this creature is exceptionally painful. Ordinarily its movements are extremely sluggish but upon the slightest sign of danger appearing it vanishes with the most extraordinary rapidity—the eye quite failing to follow it, so that even the slight and quite inadequate consolation of putting it to a violent death is denied one. Beside the Viudas, there were Tabanos and other Moscas bravas but none approached in badness the Viuda.

After some miles of this parklike semi-woodland type of scenery we suddenly found it gave place to another type—a far reaching palm forest or palmar, open plain with tall coarse grass and dotted at short intervals with numberless tall and slender palm trees (*Copernicia cerifera*) with roundish heads of fanshaped leaves. We stopped in the middle of the palmar during siesta time, and the dead silence, the perfect stillness of everything and the dull and sombre tints of the palms, combined to produce a weird and somewhat depressing effect. Nelson and I clambered ashore and felled one of the palms, extracting the fleshy heart of its crown of leaves. With these and the steaks cut from the tail of last evening's Jacaré we managed to make a meal quite in harmony with our surroundings—not unpleasant: the flesh of the Jacaré white and fishlike in appearance; the palm core reminding one slightly of raw turnips, though boiling changes its colour from yellowish white to pale violet. In addition to the Jacaré and young palm we had gathered some guayabo or wild guava which made a fitting dessert.

In the afternoon in one of the long reaches we saw a number of animals exactly resembling seals in their manner of bobbing up and down. The Argentines called them indeed such (lobos). They were however extremely shy, keeping well ahead of us for a time and then disappearing under water and coming up again far astern so as to give one no chance of a shot. However there could be no doubt that they were really otters (*Lutra paranensis*) though so seal-like in their behaviour.

Apart from them we have seen little of special interest except perched on a tree by the river, a pair of large Muscovy ducks (*Cairina moschata*) a characteristic duck of the Chaco and the warmer parts of the continent of South America. Incidentally it provided the early Spanish explorers with the ancestors of the Muscovy duck of our farmyards.

No Jacarés at all have been visible, owing no doubt to the cold. Dull and cloudy all day with some rain; we have ourselves felt the cold acutely—not having anything in the way of clothes with us except the very thin ones on our backs.

So far we have found the river comparatively free from dangerous snags and with usually about a couple of fathoms of water; the chief danger to navigation has lain in the rapidity of the current which is running at least five or six miles an hour and causes below each bend the formation of whirlpools and cross currents which do their best to hurl us against the bank.

As there was bright moonlight we thought we might try to establish a new record in the history of the Bermejo and so after dinner got under way again. We kept in mid-stream so far as possible and for the time all went well. Then we felt the *Perseverancia* being gradually pulled over several degrees from the vertical and found we were on the edge of a large eddy. Gradually with the utmost attainable pressure of steam, we worked our way through but almost immediately were in troubled waters again. And this time, just at the most critical moment, a loud snap was heard as one of the rudder chains gave way and the *Perseverancia* now helpless, was whirled round and finally shot right across the river, cannonading into the opposite bank with great violence. Fortunately the bank was here soft and grass-covered so that no great damage was done. The current immediately swung us round and we were carried rapidly down stream until at length caught by a return eddy we gently touched a low lying part of the bank and the Chileño—one of the crew—was able to leap ashore with a hawser and make fast to a tree. Whereupon we remained for the night and decided in future to navigate by day alone.

3rd March: Long winding reaches and beautiful scenery. The vertical banks twenty to twenty-five feet in height crowned with dense forest, scattered tall trees with smaller in between and a dense undergrowth practically impenetrable. We were struck by the scarcity of flowers. The flora here was of course strictly indigenous, none of the European immigrants which had been so conspicuous in the La Plata region. Of the indigenous flora a peculiar tree fern and a huge Equisetum about eight feet high and over an inch in diameter seemed to carry one in imagination into the far back geological times of the Carboniferous period. Ani and Guira cuckoos (*Guira piririgua*) and Muscovy duck were the most conspicuous birds, while on the river bank where low lying there were tracks of innumerable Carpinchos (*Hydrochoerus capybara*). Of the last mentioned I shot one but it was not convenient to stop to recover it.

At night slept as usual on the bare deck and very soundly, rolled in my rug.

4th March: We have felt thankful for the cold these last two nights for it has driven away the mosquitoes almost completely during the later hours of the night.

Before casting off our moorings several familiar birds were seen—a flock of Urracas, numbers of Anis, and one or two Muscovy ducks. Besides these many flocks of green Parakeets flew over our heads—always methodically sorted out into pairs.

About eight o'clock the first of the Argentine forts on the Bermejo was passed—Fortin Bosch. However it seemed quite deserted and our siren attracted only a party of armed Paraguayans who had penetrated thus far in their clumsy boat to cut wood. So we kept on our way. A large deer floated past us and we saw a little cavy darting along the bank.

All this while the country bordering the river has been becoming more and more densely forest-clad—the intervening stretches of grassland becoming smaller and less frequent. The banks on either side are steep or vertical, 20–30 feet in height and crowned with luxuriant semi-tropical forest. The general effect of this is dark and gloomy from the large proportion of Nectandras and other trees with laurel-like leaves. Here and there a little variety is lent by a feathery Mimosa, or a huge candelabra-like Cereus or a Pindó palm with its stout straight stem and long gracefully drooping pinnate leaves.

Across the stream flitted with wavy flight beautiful Kingfishers in plumage of grey and red (*Ceryle torquata*) and snow white egrets; an occasional large grey heron (*Ardea cocoi*) would rise with slow flight uttering its sharp metallic cry of co-coi. Storks were seen pretty frequently, while a smaller species (*Tantalus loculator*) the so-called Wood Ibis, from the curved bill resembling that of an Ibis—was seen in immense numbers soaring at a height so great that they appeared merely as tiny specks in the sky.

Towards 10 o'clock, some 72 miles from the mouth of the Bermejo we reached Ñacurutú Island, about 400 yards in length, triangular in shape, its base formed of hard tosca facing up stream and parting the waters into a narrow channel on each side. With its precipitous banks crowned by dark and gloomy forest and the wildly rushing waters between, the river here makes a striking bit of scenery. Just above the island we passed a rough dug-out canoe tied to the bank—the first sign of wild Indians. The owners themselves were unseen though no doubt watching us all the while. Against the blue sky overhead soared an immense flock of Wood Ibises in ceaseless circles—attracted perhaps by some fire in the camp. And now we reached our destination Puerto Expedición—another of the Bermejo's chain of Forts. It also, like Fortin Bosch, had had its garrison withdrawn; only a small guard remaining—sufficient with the few rancheros living close by to beat off any ordinary attack. The sound of our siren soon brought out some of the inhabitants and we noted that here everyone constantly carried his Remington over his shoulder. The place consists of a collection of small mud ranchos with thick walls pierced with tiny loophole-like windows and is of course situated in the middle of a wide expanse of open grassland—it being a rule in the Chaco to keep military camps and posts well away from forest affording concealment to Indian enemies.

The afternoon I spent collecting—especially plants and insects—birds being hardly visible at all. Higginson and I went off together and the heat proved very trying so that we were much elated at coming across an old plantation of melons which we hastened to plunder.

5th March: was spent at Puerto Expedición; the *General Paz* arriving in the morning: the *Bolivia* a little later when I lost no time in getting back to my own quarters. In the afternoon we made a short reconnaissance up stream—Zorilla coming on board with us. However nothing special was seen. On the mud by the river margin were numerous trails of jaguar, some quite fresh, but we didn't see any of the animals themselves. In the trees overhanging the river we found a flock of Yacu hu (*Penelope obscura*), birds about the size of a pheasant which with their congeners occupy here the place of the Pheasants of the old World. They were very tame and shooting them could hardly be termed sport as after each discharge the survivors did not fly away but merely hopped from one branch to another with an expression of mild surprise as their companions dropped to the ground. We landed some distance up stream on the left bank of the river and going up a slight rise in the ground had a clear view over a considerable distance. No Indian smoke was however visible. While so engaged I suddenly felt as if hundreds of red hot pins had been thrust into my body and looking down observed to my horror legions of large red ants swarming up my trouser legs. I had in fact taken up my position right over an ants' nest. Dropping my rifle clear of it I fled in all haste to the river, jumped in and tore off my clothes and then pulled off the horrid creatures one by one. By this time most of those in my discarded garments had forsaken them on finding their prey gone, so I was able to reclothe myself, pick up my rifle, and get back to the *Bolivia*.

6th March: This morning Page gave me the rather startling information that I was to take the wheel and have the responsibility of navigating our ship down the flooded waters of the Bermejo to the Paraguay. I had naturally misgivings, the *Bolivia* being a much more tender boat than the *Heave-up*, and I knew that for her to touch a projecting mass of tosca or a submerged tree trunk would mean the end of everything. However orders were orders.

We made a start shortly after midday and the run was an exciting one, for owing to the great rapidity of the stream and the numerous cross currents I had to cram on every ounce of steam and keep going at the topmost limit of our speed in order to have as complete steerage control as possible. So absorbing a task was it that at its end I was quite unconscious of anything we had passed on the banks—including several Jaguars and a Tapir. Only one or two incidents stood out in my consciousness, how at one point the current had gradually worked us in close to the hard tosca bank and how we flew along at 15 knots or so within a few feet of the bank, not daring to put the helm hard over which would have made the stern swing inwards and involve certain destruction, Page all the while perfectly motionless by my side, not saying a single word, and how gradually, very gradually, it was possible to edge out from the bank towards the centre of the stream: and how again at another point our port wheel just grazed a bank and had the corners of several floats ripped clean off as if cut with a knife. At another time there was visible ahead of us the telegraph line to Formosa—the military capital of the Central Chaco. It sagged down in the middle of the stream where it was clearly impossible to pass clear underneath it so I had to risk keeping close in to the right bank where the wire was at a greater height above the water. However even so we did not clear it: the tip of the funnel caught the wire and at once carried it away, thereby severing for an indefinite period direct telegraphic communication between Formosa and the capital of the Chaco territory.

Finally we reached the mouth of the Bermejo and I was glad to hand over the helm and have a cup of tea with Page. It had taken us six and a half hours to traverse the distance which in our upward journey in the *Perseverancia* had occupied three days. Emerging into the Rio Paraguay we turned South and reached Timbó shortly after dark. Presently the *General Paz* arrived, with her fore part on one side battered in and crumpled as if made of paper. She had touched twice—fortunately on both occasions on soft ground. It was a splendid tribute to the Clydeside workman-snip that not a rivet had started and hardly a drop of water had leaked in through the injured part.

That journey down the Bermejo was a considerable strain and I have seldom felt more tired than when at last we slid out from the obstreperous Bermejo on to the placid waters of the Paraguay.

8th March: After a day of digestive upset and considerable fever at Timbó I was again able to resume my collecting in that neighbourhood. In the morning I took my former path to the Northward parallel to the river and after about half a mile turned inland through a piece of open camp with a small laguna and patches of rush-grown swamp, and had there a successful morning's work. The most interesting new bird was *Ardea sibilatrix*, a small heron with a clear hard whistle recalling the metallic note of the Bandurria (*Harpiprion caerulescens*). Its flight also differed from that of its congeners, the wings being moved through a very small vertical angle and with much greater rapidity than usual. On my approach it adopted the camouflage attitude seen in various other species of heron, standing perfectly motionless with its back towards me and its neck and head extended vertically upwards. The colouring of its back merged so closely into that of the vegetation that it was almost invisible.

At the village post office we had some trouble about a letter addressed to England. The postmaster searched in vain in his list for the name Inglaterra. Europa was there, España, Francia, and many other countries but no Inglaterra. He indeed seemed sceptical as to the existence of such a country but at last agreed to accept our assurance that it was a part of Europe. In the evening we left Timbó en route for the Pilcomayo and steamed on till late.

9th March: Next day our progress was interrupted by a call at Villa Pilár, the fairest spot we had so far seen. Just below the town were rugged islands covered with beautiful woods. The town itself of charming cottages, brightly painted or whitewashed, with low red roofs and wide verandas all round. Over all floated the tricolour flag of Paraguay—pleasing to the eye after having seen nothing but the blue and white of Argentina all these past weeks. The weather was fine and sunny but the air cool and balmy and we now came in contact for the first time with the inhabitants of the Lotus-eating land of Paraguay. The people clustered lazily about the point where we were moored, to watch the unaccustomed spectacle. We much desired some fresh fruit. They told us where we could get plenty but were quite unmoved by our offers of high financial reward for fetching some, or even for merely going to take word of our desires to those who had the fruit to dispose of.

10*th March:* The Paraguayan shore like a garden. Passed Formosa. On a sandbank were a flock of the great stork called Jabirú (*Mycteria americana*), spaced out widely apart as unsociably as usual, each standing on one leg its long black neck folded up on its shoulders.

11*th March:* Cast off and under way about 5 a.m. and on its becoming light we found ourselves about 6 leagues below Asunción the capital of Paraguay. On the West bank of the river there was for a time open palmar like that of the Bermejo. There was much camelote floating on the surface of the water which hereabout showed a pretty uniform temperature of about 80° Fahr. Many Kingfishers—the beautiful Ringed Kingfisher (*Ceryle torquata*) the largest of the South American species, and also the smallest (*C. americana*), a good many Jacarés and a few Carpinchos. We paused at Puerto Pilcomayo the Argentine frontier post marking the mouth of the Pilcomayo and there we left the *General Paz* with the troops, and pushed on towards Asunción.

The Eastern or Paraguayan bank was particularly beautiful. The dark clear water swirled past the high cliffs of Itapytapunta —formed of soft red sandstone with a cave here and there and mantled with rich vegetation, its deep green contrasting vividly with the red of the cliff [Plate II (*a*)]. At one point a charming little Paraguayan cottage with its low red tiled roof and wide veranda nestled in a little valley sloping down to the water's edge and embowered in a luxuriant growth of bananas, guavas and palms. I thought it the most enchanting spot I had seen and dreamed of settling down there to a lotus-eating existence safely shut away from the European world and all its complications.

About 3 o'clock we arrived at Asunción and I immediately went ashore to pay my respects to the British Consul, Dr William Stewart, a very agreeable fellow Scot who talked interestingly about Paraguay and its natural history and promised to give every help in his power towards the success of my stay in South America.

We are to remain a couple of days in Asunción which will give me time to get the spirit tanks fitted and also to obtain various needed stores and odds and ends of equipment.

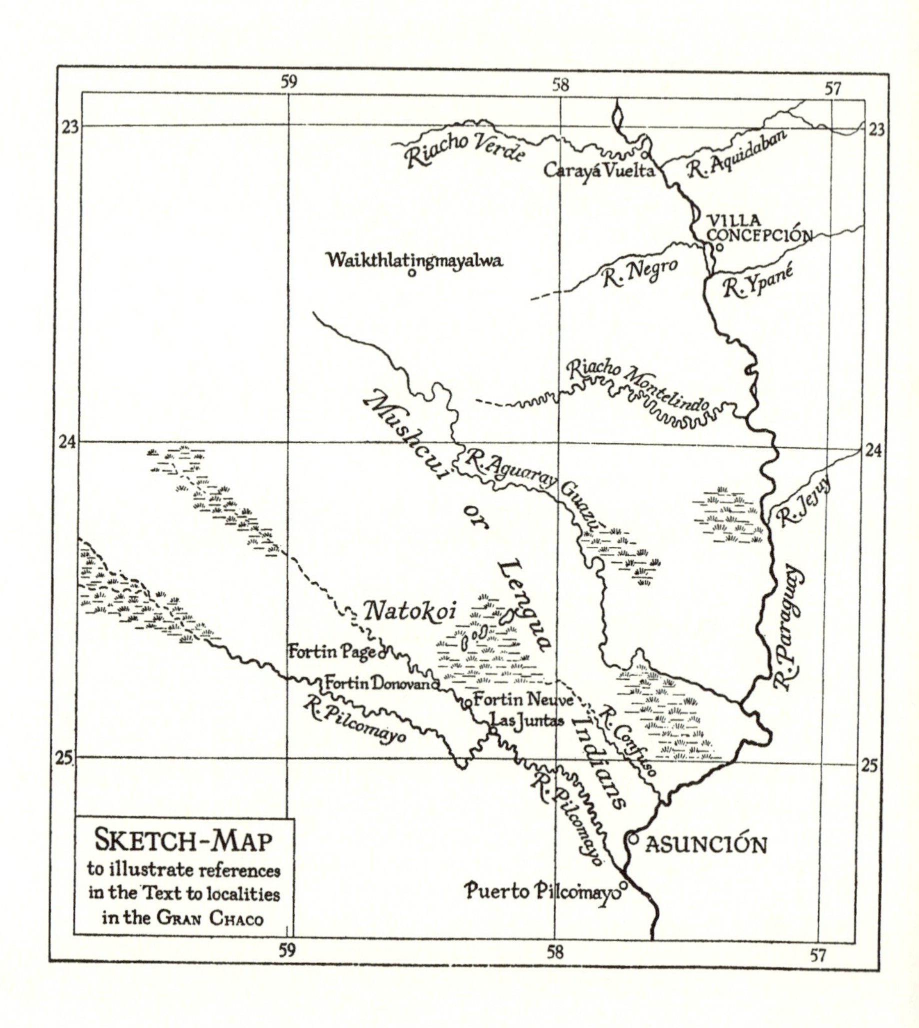

59
58
57
23
24
25
Riacho Verde
Carayá Vuelta
R. Aquidaban
VILLA CONCEPCIÓN
Waikthlatingmayalwa
R. Negro
R. Ypané
Riacho Montelindo
Mushcui or Lengua Indians
R. Aguaray Guazu
R. Jejuy
R. Paraguay
Natokoi
Fortin Page
Fortin Donovan
Fortin Neuve
Las Juntas
R. Pilcomayo
R. Confuso
R. Pilcomayo
ASUNCIÓN
Puerto Pilcomayo
SKETCH-MAP
to illustrate references
in the Text to localities
in the GRAN CHACO

CHAPTER V

THE RIVER PILCOMAYO

12th March: About 6.30 a.m. I went ashore with Pool to see the much heard of market place. We found it to be a large roofed structure filled with a motley crowd of buyers and sellers. The stalls were almost entirely kept by women, many of them good-looking, and occasional ones here and there exceedingly handsome. One was a real beauty—a girl of about seventeen—slender in figure, luxuriant fair hair, faultless features, lips just sufficiently pouting to give piquancy to her beauty. Her complexion was fair, with rosy flush suffusing her cheeks, and her general air of real distinction was not in the least detracted from by the shawl over her shoulders and her bare feet. With that complexion and those liquid blue eyes, one might well have taken her for a Lancashire Witch rather than a native of the land of Paraguay.

Having paid our tribute of admiration to the Paraguayas we settled down to purchase various delicacies, had a delightful cup of coffee, and returned to the ship.

Here a double shock awaited me. Hansen, one of the crew who had gone ashore last evening had borrowed my Webley R.I.C. revolver and had failed to return: and a second and far worse shock was the announcement by the Commander that we should leave Asunción almost at once. I had just time to dash to the chief gunsmith's and purchase what purported to be a Webley R.I.C. revolver of the same type as that with which Hansen had absconded, but which turned out to be a German imitation duly ornamented with the Webley trade-mark but most inferior in quality.

We steamed off at about 9.30 a.m. and after a rapid run down-stream and a short pause at the Argentine frontier post, we found ourselves entering the Rio Pilcomayo.

Once within the river the ship's company were called together and addressed by the Commander. We were from now on to be under strict military law, and were warned of the penalties that infractions would bring. Certain faults would automatically be punished with death—amongst them the theft, even on the smallest scale, of provisions. Many gentle people no doubt, would regard the death penalty for stealing a biscuit as shockingly cruel: in fact it was nothing of the kind. The death penalty without appeal works out in practice as an effective deterrent. The biscuit is not stolen!

One of the details that had to be arranged was that of regular watchkeeping during the hours of darkness and I was glad that the others agreed that our two hour watches should remain unchanged instead of varying from day to day as is customary at sea. My own watch continued to be from 2 to 4 a.m. and we all found ourselves easily accustomed to turning out regularly at the particular hour of the night when our watch began.

I had been surprised by the small size of the Bermejo but the Pilcomayo turned out to be even smaller—only about 60 yards in width. The water, averaging 3–4 faths. in depth over a bottom of sandy mud, was clear and dark; pleasant to drink though with a faint suspicion of salinity. The current at this time was extremely slow, not more than a quarter of a mile an hour, owing no doubt to the waters being dammed back by those of the Paraguay, now in a state of high flood. The banks high and steep, and covered everywhere by a growth of rank vegetation—coarse grass, Tacuara canes with their fanlike arrangement of leaves, creepers of all kinds, and patches of forest. Trees fairly continuous along the bank but in many places the narrow belt of trees backed by open campo. Flowers much more conspicuous than on the Bermejo—more especially a dark pink Convolvulus. Of animal life we saw several carayás and carpinchos: and many otters. Also a good many jacarés. Of birds there was quite a good variety. By far the most abundant was the black Ani cuckoo which as we steamed along flew up in hosts on either side, uttering their shrill piping cries. Next, after them in abundance were the two species of Kingfishers already mentioned. Cormorants and Darters (*Plotus anhinga*), Little Bitterns, Green Parakeets, Cardinal finches and Sappho Hummingbirds made up a fine show of bird life.

In this lower part of the river we found navigation easy—plenty of water, and snags few and far between. The chief trouble came from the tortuosity of the river and the sharpness of the curves, the *General Paz* in particular rocketing into the bank at nearly every bend while even the shorter *Bolivia* did so every now and then.

13th March: In the morning we arrived at the last outpost of civilisation, the estancia of Don Pedro V. Gil, and remained there for the rest of the day, taking on board a supply of beef. This was preserved in the form of charqui, dried in the sun.

After this simple process of 'dehydration' the beef will remain fresh for prolonged periods so long as care is taken to keep it dry. It makes an excellent meal when broken up by being pounded with a wooden club and given prolonged cooking, with added vegetables such as rice or pumpkin when available.

After breakfast I went ashore with my gun but had again forced upon me the uselessness of going bird-collecting in the heat of the day, as barring a few black vultures (*Cathartes atratus*) not a bird was to be seen. *The* time for collecting birds is about, and for an hour or two after, sunrise. Again in the evening birds are to be seen pretty numerous but during the heat of the day they go to roost. The afternoon on the other hand was clearly the time to entomologise or botanise. The whole air resounded with the hum of insect life. Jumping about in the grass were what seemed an endless variety of Grasshoppers while above flew dragonflies, again in great variety. The dragonflies commonly were to be seen perched on the top of a stick, turning about their great heads—almost entirely made up of eye—with quick jerky movements, keeping a look out for their insect prey in pursuit of which they would every now and then make a rapid dart. As I walked along with a cloud of mosquitoes hovering round my head there would often be one or two dragonflies in attendance hunting the mosquitoes. The jerking movements of the head in a stationary dragonfly have to do with the fact that the type of eye met with in such insects is adapted for the detection of objects in relative movement, rather than the formation of a detailed picture as in the 'camera' type of eye found in man and other vertebrates. There were many and beautiful butterflies, amongst them a magnificent Swallowtail.

Of birds the two common Cuckoos, the Blue Jay (*Cyanocorax chrysops*), a little chocolate coloured dove (*Chamaepelia talpacoti*), Ovenbirds and a Trogon were the most noteworthy.

14th March: In the morning I put in some more time bird-hunting and got a good bag, including several species not hitherto recorded from the Argentine. Among these was a strange chestnut-coloured Cuckoo (*Piaya cayana*) with an enormously long tail which I found feeding on the ground in a patch of thick forest. One of the things by the way that particularly strikes the European naturalist is the extraordinary variety of the Cuckoos (Cuculidae) in this part of the world. They are also in great numbers, recalling the groups of crows (Corvidae) of the old world which are here represented only by two species of Jay.

The Estancia Gil consisted for the most part of rich open

parklike pastureland bearing a luxuriant growth of fine grass and varied by patches of forest. I noticed however, plentifully scattered about, shells of the large water-snail *Ampullaria*, indicating the liability to flooding during the wet season. The cattle were large and in good condition and the beef we found excellent in quality. Round the estancia house was a stockade about 15 ft. high which however had been allowed to fall into disrepair as there had been no Indian raids for several years.

At 10.10 a.m. we cast off our moorings and steamed on our way, leaving behind the last civilised settlement we should see before reaching Bolivia.

15th March: Ashore in the forest while trees were being felled and chopped into logs for fuel. Under the tall trees grew others small and slender, including one called Arrayán or in Guaraní Ñanga piru, from the leaves of which is made an infusion used as tea. The ground was littered with prostrate tree trunks in all stages of decay on which grew fine ferns. Otherwise the ground was covered with a peculiar soft grass and it was easy to move about. There being little undergrowth I was able to wander about freely in the forest, always taking care to blaze a tree at short intervals so as to enable me to retrace my steps—a necessary precaution—the complicated turns and twists of the river making the compass a quite insufficient guide. On many of the trees were spherical Termite nests up to about three feet in diameter built of wood particles and each provided with a covered way running towards it up the side of the tree trunk. In one of the rotten prostrate trunks was a nest of large ants, black except the prothorax and abdomen which were covered with golden bronze hairs. The workers were about half an inch in length and stoutly built while the soldiers were larger, ¾ of an inch, and provided with enormous jaws. As I proceeded slowly through the forest I heard subdued mutterings and saw my dog Turco gazing with interest up into one of the trees and looking up there were visible small marmoset-like visages with huge owl-like eyes. They belonged to the rare Mirikina monkey (*Nyctipithecus trivirgatus*) adapted by their large eyes to nocturnal habits but yet showing no signs of being blinded by the daylight as Azara describes. They were beautiful little creatures, grey above and reddish yellow underneath, and it was hateful to have to sacrifice them to the cause of science.

In the forest were many Jays, of both species, Azure (*Cyanocorax caeruleus*) and Urraca (*C. chrysops*). The former has been described as extremely wild and difficult to shoot but here they were not in the least shy and hopped about in the branches watching me inquisitively, all the while keeping up a concert of shrill pipes or, when particularly close, of harsh screams.

An interesting acquisition was a Woodpecker new to science and named by Hargitt after its discoverer *Celeus kerri*, shabby in appearance with faded-looking rusty brown plumage smelling strongly of musk, and its pale yellow headpiece always looking soiled and bedraggled, in striking contrast with the brilliant scarlet cowl of some of the other large woodpeckers of the Chaco.

Another great sight was a specimen of the Mitu (*Crax sclateri*), a magnificent Curassow as big as a Turkey, perched among the top branches of a tree by the riverside. Nelson and I landed in pursuit but failed to secure it.

Now that we were well into the Pilcomayo our progress settled down to a daily routine. Starting at first daylight we steamed upstream at constantly varying speed. Progress was of course very slow, the engines being stopped every few minutes owing to the sudden bends in the river at each of which we ran violently into the outer bank. On each side of bow and stern were stationed two or three men with large poles or botadors with which to shove off when such contingencies occurred. All the while a chart of the river was being constructed, each reach being mapped according to its compass-bearing and length. The mode of measuring the length of a reach was such as would shock a modern cartographer. Our commander was a dead shot with the rifle and he would aim at some point at the far end of the reach and by adjusting the sight until his aim was accurate he so determined the distance!

When fuel was exhausted we would tie up to the bank and the next few hours would be occupied with felling suitable trees and splitting them up into logs of the proper size for fuel. This task finished we passed the logs on board, swinging them from hand to hand. On a hot afternoon this monotonous work was extraordinarily soporific and I would often fall asleep for a fraction of a second, waking however before I had time to fall down. These logs rich in resin were magnificent fuel, and stoking the furnace was one of the hardest bits of work I encountered, each log seeming to vanish at once when thrown into the furnace. What made the work particularly trying was the absence of forced draught which in the case of the *General Paz* and *Caa Guazú* kept the stokehold moderately cool.

It was in the late afternoon towards sundown that I had my most interesting time. Taking with me rifle and shotgun I would go off alone in the canoe paddling on ahead of the ship as noiselessly as possible. Sheltered by its high banks the river surface was a perfect mirror reflecting the forest-clad banks. Troops of Ani cuckoos uttering their shrill piping cries, pairs of Blue-fronted Amazon Parrots (*Chrysotis aestiva*) keeping up a constant and no doubt loving connubial conversation, an occasional Toco Toucan; and the two common kingfishers now joined by a third (*C. amazona*) intermediate in size. One time a large tree crowded with snowy egrets, scores of them, rosy in the rays of the setting sun, made a lovely picture. On another tree-branch projecting over the water sat an enormous Ñacurutú Owl—a variety of the Virginian Owl (*Bubo virginianus*)—with a peculiarly deep sepulchral voice.

At one point a little creek overhung with branches opened into the river and pushing the canoe through it I found it widened out into a laguna bounded all round by dense forest and much of its surface covered by a floating carpet, here of Camelote (*Pontederia*), there of the smaller *Pistia*.

In other places were the enormous flat leaves—several feet across—of the Victoria water-lily (Plate II (*b*)), and there one saw a charming sight, a troop of Jaçaná (*Parra jacana*) slightly built birds allied to the plovers, chestnut above and black underneath, carrying a yellow spine on each wing and possessing enormously long toes with long straight claws—an adaptation to enable the bird to support itself on the treacherous carpet of floating water plants. There the jaçanás paced about, picking up particles of food, and suddenly as if at a given signal the whole troop would stop their feeding and stand motionless, holding up their beautiful lemon yellow wings vertically in the air, and then suddenly as if again at a given signal fold them up and proceed with their feeding. In the open water were to be seen what looked like partially submerged floating logs, each however having the peculiarity of showing three small knobs projecting above the surface, and the further peculiarity of slowly subsiding out of sight as the canoe came into its neighbourhood. They were, in fact, not logs but jacarés, and sometimes if a group of ducks swam over the point where a jacaré had submerged one of them would be seen to disappear suddenly under water with violent struggles as the jacaré dragged it down by the feet. Even more formidable than the jacaré was a pestilential fish called the Palometa (*Serrasalmo spilopleura*) whose knife-like teeth enable it to bite out solid chunks of the flesh of a man or other mammal that falls into the water. Fearsome tales are told of the way in which a man or bullock can be reduced to a bare skeleton in an incredibly short time after falling into the water. In the side lagunas of the Pilcomayo these palometas were abundant and one could often see their dorsal fins ploughing along the surface of the water.

Returning down river in the deepening dusk one might get a glimpse of some of the larger mammals. A family party of absurd-looking carpinchos would plunge into the water as one approached: a jaguar, accompanied by its kitten, putting its tail against a tree with quivering tip exactly like a domestic cat: or a tapir—shyest of all the Chaco mammals, visible just for a moment as it crashed away through the vegetation.

16th March: We passed the mouth of a tributary—the Rio Negro—with clear dark water opening into the Pilcomayo on its left bank. Next day we encountered for the first time a species of Curassow—the Charata or Yacu Caraguatá (*Ortalis canicollis*)—which at later stages was to be one of our chief food birds; brownish in colour and with a heavy flight something like that of a pheasant. This species turned out to be abundant in all the thick forests of the Pilcomayo. They occasionally descend to the ground to feed but usually remain among the upper branches of the trees, feeding on various fruits. They are sociable birds, many being usually found in company. They are very timid, and yet very curious. When one enters a bit of forest in which the charatas are unaccustomed to the sight of man they examine the intruder curiously and summon their companions with their soft and cheepy call-notes. If one remains perfectly still all the birds within hearing collect round and by answering their call-note one can bring them down to within a few feet. If one walks towards a tree in which are some charatas they first utter their soft call-note to draw their companions' attention and then begin to hop about uneasily, their voices rising in pitch by regular gradations until they end in shrill screams amusingly expressive of fear and timidity.

At times, more especially about sunrise, the community of Charatas unite together to produce an extraordinary din. They utter loud and harsh cries, something like the sound of a gigantic rattle or of the syllables chacarata, chacarata, from which they get their name. All the birds in one part of the forest uniting together the effect is almost deafening. Other companies of birds answer and on a fine morning in the Chaco just after sun-rise one hears these Charata choruses resounding in all directions.

Of ducks, the commonest here is the Muscovy duck (*Cairina moschata*) of which to-day we have seen many broods nearly fully grown.

In the forest were many little pools, most of them covered by a floating carpet of *Pistia* upon which stalked about lovely Jaçanás —the adults with their plumage of chestnut above, black breast and pale lemon-coloured wings and the immature birds with dark bronze plumage on their upper side and white underneath.

The most striking bird was, however, the Urraca Jay (*Cyanocorax chrysops*) conspicuous in its beautiful plumage of blue and yellow, and still more so from its boldness and curiosity which caused one often to be surrounded by a mob of these birds, some approaching to within a yard or two, all eyeing one curiously and keeping up all the while a chorus of cries of every imaginable pitch, from a shrill scream through a flute-like pipe down to a deep hoarse bass. In many cases the Urraca Jay was accompanied by the Azure Jay (*C. caeruleus*) quite similar in habits and disposition but with a voice apparently limited to a harsh scream.

A curious wailing sound seemed to come from the depths of the forest. As I slowly made my way in its direction the sound grew in volume and became a chorus of weird and mournful cries as of children in pain. Then I came to a forest pool, dark and sombre, its waters partially covered by a floating carpet of *Pistia* shut in on all sides by the gloomy forest wall, with a group of white egrets perched motionless on an overhanging tree, and the cries suddenly ceased—heavy flops into the water telling that the vocalists had been frogs, in all probability Paludicola [see p. 195].

19*th March:* Last night a weird experience during my watch, the normal black darkness of the forest being replaced by a pale greenish phosphorescent light—due no doubt to a photogenic bacterium or fungus growing on the surface of the decaying leaves which covered the ground everywhere.

The *General Paz* has been giving great trouble owing to her length making it almost impossible to negotiate the sharp bends of the river. So it has been decided to send her off downstream under the command of Lieut. Mate and this morning our ship's company on the *Bolivia* was increased by the arrival on board of Commander Zorilla along with Col. Racedo in command of the troops with two junior officers and 25 men. There also came the Doctor of the Expedition, Luigi Vignoli. The *Bolivia* was now grossly overcrowded and the whole vessel throbbed when under way with the unwonted strain on the engine. However the military party will sleep and feed ashore, and when the country allows it march overland.

In the afternoon among birds shot were a couple of Curassows —one the big Turkey-like Mitu and the other a Yacu hu—and a number of Muscovy ducks which included one magnificent bird measuring no less than 35 inches in total length. A large jaguar swimming across the river just ahead of us escaped into the forest though wounded.

21st March: The country now is very much open palmar with luxuriant waist-high grass. About 3 o'clock we steamed into the 'Laguna de las palmas' about 400 yards by 250, with 1–3 faths. of water and bounded by a particularly dense growth of palms. Returning to the river about an hour's steaming brought us to Las Juntas where the river divided into two branches [Plate III (*a*)]. Here we found affixed to a tree an inscription recording the passage four weeks earlier of a surveying party under O. J. Storm of the Argentine Department of Hydrography.

The branch to our left appeared considerably larger—its current, however, was much less and its water quite salt. Moreover, the River Pilcomayo having been accepted as forming the frontier between the Argentine and the Paraguayan Chaco it seemed entirely suitable that we, flying the flag of Argentina, should regard the Northern branch as *the* Pilcomayo. We accordingly pushed our way into the right-hand channel. However, progress from Las Juntas onwards was very slow, the river being exceedingly tortuous, very narrow—only 15–20 yards in width, and overhung by trees which often stretched right across the channel and had to be cut away to allow of our passage. Where there were none of these great tree trunks barring the way it was possible to steam slowly onwards, our high upper works brushing against the branches and our decks becoming littered with a thick layer of lichen-covered bits of twig. This led to a fascinating discovery, for movements here and there amongst the debris led to closer investigation and that in turn to the discovery that a large part of the debris consisted not of lichen covered twigs but of animals of many kinds all of them camouflaged to look exactly like lichen. There were snakes, lizards, tree-frogs, spiders, scorpions, beetles, mantids, locusts, ants—and we thus learned of the existence of a hitherto unknown fauna inhabiting the upper branches of the forest trees and characterised by the marvellous adaptive resemblance of its members to the lichens amongst which they lived. [It is one of the grievous memories of the Pilcomayo expedition that the loss of the main collections involved the loss of hundreds of specimens illustrating this lichen fauna.]

Conspicuous in this part of the river were two species of Hangnests—those Icterids which weave long pendulous nests suspended from the branches of trees. The one species (*Amblycercus solitarius*) forms its nest of coarse brown fibres and prefers the tips of branches overhanging the water. Its song is sweet though short, consisting of three notes—*whee-hoo-ah*, clear and metallic and remarkably rich in tone. It also frequently utters a peculiar miauw-like sound, either alone or followed by a couple of abrupt, deep and peculiarly liquid notes—*miauw-cu-cu*. The other species (*Cassicus albirostris*) is equally common and characteristic and is frequently to be seen in the tops of the Carandai palms. But though it feeds in the open its real home is in the recesses of the monte duro or hardwood forest where its deep rich notes strike upon the ear and its long flask-like nest, in this case woven of black fibres almost as fine as horsehair, may be seen suspended from the trees.

In this region the great Toco Toucan (*Rhamphastos toco*) was common and could be seen perched in the upper branches of tall trees, uttering hoarse cries, or winging its way across the river with heavy flight its great bill stretched out in front after the manner of a stork. Its coloration is striking: plumage black for the most part, with a large gorget of very pale yellow. The bill is coloured in yellow, carmine and black: the eyelids are bright blue, set in a circle of brilliant orange: the feet light blue and claws dark blue. The bill when in active use as a pair of forceps, picking little fruits or seeds, looks absurdly out of proportion and over-weighted but in fact its bony framework in the form of an open sponge-work combines great strength and rigidity with the minimum of weight.

26th March: The river is still extremely tortuous, describing curves nearly completely closed, only a few feet sometimes preventing the water from short circuiting and cutting off the current from the main part of the curve. Eventually the narrow neck of land in such cases is severed and the current passes straight through. A result of the constant changes in the river course so caused is seen in numerous dry shallow valleys flanking the main channel and each representing a deserted portion of river bed.

In the rich vegetation of the river bank three types of creeper are at present conspicuous—a Convolvulus with pale purple flowers, an Asclepiad with white flowers, and a Passion flower in full fruit, many of the trees being festooned with its climbing stems from which depend the elongated ellipsoidal fruits of a rich golden colour. In the trees too [Plate III (*b*)] are many epiphytes—especially several species of *Tillandsia*. One of these Flores del Aire, as the Spaniards call them, turned out to be a species new to science (*T. tomentosa*). Another interesting and abundant epiphyte was the Old Man's Beard (*Usnea barbata*—a lichen).

Fish are abundant in the river—the fish population being concentrated by the dry season shrinkage of the waters—and all of them eatable—the two best being the Dorado and the Pacu.

29th March: Ashore, with my dog Turco, collecting while we stopped to cut wood. The thick grassy jungle here was traversed in all directions by Tapir paths testifying to the abundance of this animal, so shy and rarely seen. After the wood was got on board the whole day was spent steaming ahead, crushing our way through the branches of trees everywhere overhanging the river and causing the usual shower of their animal inhabitants down on to our deck. In the afternoon the smoke of a great camp fire was visible some distance off and later we passed some palms evidently chopped by the savages: so we were now clearly in Indian territory.

31st March: To-day I had one of the most striking demonstrations of Nature's skill as an obliterative colourist. I happened to shoot down five out of a dense flock of six parakeets (*Pyrrhura vittata*). I could not find them although I had marked the exact spot where they had fallen: so I felt for them with my hands and soon found them lying right under my eyes though I had failed to see them, in spite of my eyesight being then both sharp and highly trained in the detection of animals.

I also got a male *Celeus kerri*—differing from the female, which I had already obtained, in the presence of a scarlet malar patch.

1st April: As a result of the dry weather our progress has been still further impeded by the gradual shrinkage in the volume of water in the river and to-day we were brought to a full stop, so it was decided to lighten the ship by putting most of the soldiers ashore, only Racedo, Lt. Donovan, the Doctor, and half a dozen troopers, remaining on board. This lessened our draught by about 3 inches and we were able to press on. The soldiers left behind were to construct a stockade fort—Fortin Regimiento

Nueve de Caballeria—and then travel overland until they rejoined us.

2nd April: A characteristic bird in these regions is the large Rail called Paca-á by the Paraguayans (*Aramides ypecaha*), with dark plumage and long red legs. It is often to be seen feeding and picking its way along the water's edge carrying its tail cocked upwards and giving it a sharp jerk every few paces. The paca-á is one of the shiest birds of the river, keeping a constant vigilant look out, and on the appearance of a stranger at once slipping into the shelter of the vegetation through which it rapidly makes its way without as much as moving a blade of grass to betray its whereabouts. It only takes to flight if almost trodden upon and it is occasionally to be seen perched in a tree. The ordinary call-note is a soft 'cluck' but during the evening and occasionally at other times several birds collect together and produce an extraordinary noise as if carrying on a conversation in loud shrieks, varied from time to time by all uniting to cry *oh-whauchee, oh-whauchee* in an endless variety of tone and expression.

The paca-á we decided to be the best tasted of all the food-birds of the region.

3rd April: One of the finest mornings we have had in the Chaco, the sun glinting against the tops of the palm trees, the sky of the deepest blue except where speckled by a wisp of cirrus: the air soft and balmy while yet exhilarating, and filled with the sounds of bird-life—the wild cry of the Chajá, the rattle-like accents of the Charatas, the clear pipe of the Urraca, and the varied chirping of the smaller Passerine birds—forming a harmonious medley delightful to the ear of a naturalist. When starting off I had an interesting experience in a different department of natural history for I saw for the first time the meaning of the orange and brown stripes on the under surface of the wings of so many of the common (Heliconid) butterflies of the Chaco, as I suddenly realised that in the chilly morning air the tall grass was peopled by thousands of butterflies hanging on to the lower side of grass stems so that the approximately vertical striping merged exactly into the light and shade of the background of predominantly vertical grass stems. Here, as in some of the fishes that swim about among the reeds and rushes, the obliterative effect depends upon harmony with the background, whereas in large animals such as zebras the obliterative effect of the striped pattern under conditions of poor visibility is due to its disruptive effect in breaking up the tell-tale continuity of the body surface.

The river is now 40–50 feet wide with a sluggish current of about one mile an hour and enclosed within permanent looking vertical banks about 10 feet above the water surface—a comment on which is however provided by a distinct watermark on the palms about 13 feet above the present water level. The volume of the water is diminishing, the average depth being now no more than 3–5 feet. For the first time on the Pilcomayo we have now encountered what might almost be called rock—a soft whitish sandstone containing much carbonaceous matter coarsely laminated and forming horizontal strata.

10*th April:* This morning I had a quaint experience during my watch. Standing ashore with the two surviving sheep grazing close by I suddenly heard a faint sound and perceived a jaguar, looking in the faint light as of gigantic dimensions, coming on in great bounds while the two sheep dashed past me one on each side. I was unarmed and all I could do was to run to meet the oncoming jaguar, clapping my hands at it. No doubt the jaguar had been interested in the sheep and my unexpected interruption made him turn off and beat a retreat, as I did in the opposite direction to the safety of the boat.

During these night watches in the forest with senses keenly attuned, one is often struck by the oppressive silence. This, on occasion, would be broken—by the sudden collapse of some long-dead forest tree or by the cry of a wild animal—and now the voices of the forest would break out in full blast—the prolonged roar of the Aguará guazú (*Canis jubatus*), the short sharp 'hu' of the jaguar five or six times repeated, the cries of various night birds. If a shower of rain had fallen, these would be heard against a background not of silence but of frog voices—catlike miauws, birdlike trills, croaks, coughs, grunts, forming together a quite extraordinary medley of sound.

I have already mentioned the wonderful lichen fauna of the tree-tops. This morning I was examining in detail one of the grasshoppers belonging to this fauna (*Aphidnia punctifrons*) and was really amazed at the perfection of its imitation of the lichen among which it lives; its colour on the upper side precisely the grey-blue of the lichen, its lower side precisely the dark brown of the under surface of the lichen. Not only so, but there extend from its body and legs little crumpled foliaceous expansions which make the imitation in form just as striking as the imitation in colour. Another allied insect, one of the Mantids, a hitherto unknown species of *Acanthops*, showed an equally striking resemblance, in this case to a decaying leaf, coloured in various shades of grey and brown and, on the abdomen, black patches imitating exactly the patches of fungal growth common on dead leaves. In another case a little bit of dry and withered leaf adhering to my shirt was just on the point of being brushed off when a tiny projecting leg betrayed the fact that it too was in reality a small Mantid.

11th April: Ashore on the southern bank and walked inland along the bed of a dried-up stream, the bare hardened mud making the going very easy compared with the usual struggling through an almost impenetrable growth of waist-high grass bound together by all kinds of creeping and twining plants. Some distance inland I found a little laguna surrounded by rushes and containing delightfully sweet water—the river water being now not merely reduced in volume but so salt as to be quite undrinkable. On the laguna were a pair of muscovy ducks, with a brood of ducklings; also a Rolland's Grebe, the first I had encountered in the Chaco. Nearby, perched on a tree, was a fine eagle, *Busarellus nigricollis*, with pale chestnut plumage and creamy yellow head. It turned out to be fairly common hereabouts, and its favourite food, curiously, the common freshwater crab, though its long sharp talons and the very rough undersurface of its feet would suggest adaptation to a purely fish diet. It utters a characteristic cry like the syllable *ĕh* much prolonged.

Near the laguna were many traces of Indians, so it is clear that we are now well into Indian territory. The Commander thinks we are near the place where a party of Bolivian troops were massacred some time ago and we are enjoined to exercise especial vigilance.

13th April: A fine morning among the birds. In a thicket of low trees I got a pair of the little green woodpecker *Dendrobates olivinus* of which there were many visible. After firing my first shot I became an object of curiosity to all the birds of the neighbourhood. The surrounding bushes seemed positively alive with little feathered creatures all shrieking their indignation. Hither and thither hopped the black and white Bushbird (*Thamnophilus major*), his fiery red iris conspicuous among the foliage. As he hops about he utters a long squeak followed by a trill-like succession of rapid chirps. Here also were many specimens of a bird I had not yet had the opportunity of observing close at hand—the Solitary Cassique (*Amblycercus solitarius*) whose pendulous nest has already been mentioned. It is about the size of our European Blackbird, with beautiful thick black plumage, and as it hops about it utters its low husky miauw-like cry. Besides these were immense numbers of all the commoner birds; the Chestnut Woodhewer (*Xiphocolaptes major*) hopping vertically up the tree trunks in diligent search for its insect food; the Little Woodhewer (*Picolaptes angustirostris*) similarly but ascending the tree trunk in a spiral instead of in a straight vertical; the two Cardinal finches (*Paroaria cucullata* and *P. capitata*) with their beautiful grey plumage and scarlet crests; the yellow-breasted Tyrant (*Pseudoleistes virescens*); a couple of other species of Woodpeckers; and little brown Spinetails.

An experience like this morning's impresses upon one the fact that the various birds of a neighbourhood collect together for the purpose of feeding, thus ensuring the thorough working over of the particular area, the insect feeders profiting by the disturbance caused by the vegetarians. The success or otherwise of a morning's bird collecting in the forest depends in great part upon whether or not one strikes the feeding ground of that particular morning, the neighbouring parts of the forest being practically deserted so that one may walk for an hour without seeing a single bird.

14th April: One of the most abundant as well as the most conspicuous local woodpeckers is *Campephilus leucopogon*—Boie's woodpecker, conspicuous to the eye by its large size, its black plumage, and its magnificent scarlet crested head; and to the ear by the prolonged roll as of a kettledrum produced by the rapid drumming with its beak against a hollow palm stem.

The ordinary language of these woodpeckers, heard when a number are in company, is a low, soft and squeaky chattering. One day I observed a courting couple, my attention being drawn by a delicate *ahem* as if to attract attention, followed by a suppressed giggle. Presently I caught sight of the performers and the sight was most entertaining. The female pecked away industriously at a tree trunk with an appearance of complete unconcern though all the while taking care that the male did not approach too near. After a while she flew off to a palm at a short distance and the male at once followed and perched on the same palm stem; cocking his beautiful head on one side, uttering innumerable *he-he-he*'s interspersed with an occasional 'ahem', all evidently charged with deep emotion. However there was no response; the female flew away again to a palm a short distance off and the performance began afresh.

It was of physiological interest by the way to notice that a woodpecker when wounded, even severely, would every now and then give an involuntary and convulsive peck—showing how deeply the pecking habit had become incorporated in the bird's neuromuscular arrangements.

18th April: As the dry season advances the river water diminishes in volume and a couple of days ago we came to a full stop and remained hard aground. So yesterday commencement was made to put into practice a bright idea of the Commander's, dependent upon (1) the slope of the Chaco plain towards the S.E. being extremely gradual and (2) the river channel being delimited for the most part by well-marked vertical banks several feet in height. The idea is to construct a series of dams, each of which would dam back the waters until deep enough to make further progress possible. There were some grumbles at the thought that each dam would constitute a barrier against our return to the outer world, but the old man was determined and would not for a moment contemplate the possibility of an eventual retreat down stream.

So yesterday a commencement was made by constructing Dam No. I with palm stems, brushwood, grass and earth; and by this morning the water level had risen about 7 inches and the *Bolivia* was triumphantly afloat.

24th April: Constructed Dam No. II.

28th April: The two species of American Vulture are now abundant; soaring round and round watching for carcasses while keeping a careful eye on one another for any symptom of interest in the ground beneath.

The undergrowth in the patches of forest hereabouts is composed in great part of the Caraguatá-i—a kind of wild pineapple the fruit of which, though attractive and sweet smelling, has a very acrid taste. The leaves are about 3 feet in length, their margins beset with sharp curved spines, some pointing upwards, some downwards so as to make the forest practically impassable. The leaves are concave above and pass downwards into sheathing bases in which there is always to be found even in the hottest, driest, weather a supply of clear cold water. This became at times our sole supply of natural drinking water for, as the river water shrank in volume it became more and more salt and undrinkable, and the same was the case with the water from the wells which we dug on the sites of dried up lagunas.

30th April: The river at this level is divided into two channels so our stoppage in one of them through lack of water has entailed our constructing two separate dams (Dams III and IV), one on each branch.

1st May: A little laguna about half a mile from the river bounded by sandy flats, mostly covered with an efflorescence of crystals of pure salt (NaCl), on which a pair of beautiful little Plovers (*Aegialitis collaris*) ran about with great rapidity, bobbing their heads up and down when they paused. When alarmed they would remain absolutely motionless and then the coloration of the upper surface harmonised in the most perfect manner with the background of sand so as to make them almost invisible. When approached too close they would rise into the air with extremely rapid flight, mounting upwards until lost to view.

The male had a sweet little song of short, very soft and melodious, whistle-like notes.

5th May: Most of the forenoon I was busy baling water out of my hold and preparing a specimen of the sulphur-headed woodpecker (*Celeus kerri*) shot yesterday. Then after breakfast Kennion and I went off in search of game—he taking his cylinder 12-bore and ball cartridge, and I my express. In approaching the open flats we cautiously peeped round the corner of the wood and there were visible a troop of half a dozen wild pigs. The wild pigs of the new world are a quite distinct group from those of the old, more primitive in structure, and known as Peccaries (*Dicotyles*). There are two species and these belonged to the smaller and less formidable, distinguished by a light coloured mark round the neck (Collared Peccary, *D. torquatus*) [Plate IV]. As they were rather far off we ran rapidly towards them and at about 80 yards I fired my left barrel at the largest and thought I had hit it, but it dashed off and disappeared among the vegetation. Then making a reconnaissance in the direction of the other five peccaries we raised another which dashed off towards the edge of the monte over the open ground and turned a somersault as my bullet ploughed along the right side of his spine. We then proceeded to follow up the blood-marked trail of the first peccary and presently found her lying dead just inside the margin of the grass. The heart was practically destroyed by the express bullet. It is very noticeable that an animal hit in the heart makes a wild dash forwards, or if a bird rockets up into the air, in contrast to a hit in the brain which brings all muscular movement instantly to a dead stop.

7th May: At last—after a week's wait—a sufficiency of water has accumulated above Dams III and IV and we are able to steam onwards. The country is now all open palmar with no ordinary trees except along the river margin. An unexpected mammal has turned up, a small rabbit apparently a variety of *Lepus brasiliensis*; unexpected because the particular group of rodents which includes the rabbits and hares is practically non-existent in S. America, where their place is taken by the great variety of cavies—from the Carpincho—as big as a sheep—down to the small wild guinea-pig.

10th May: Having again come to a full stop owing to the shallowness of the water, construction was started on Dam V. As provisions are running very low a Sergeant and two men were sent off downstream to hurry up the canoes which are supposed to be bringing up supplies from the *General Paz* left behind on the 19th March owing to the impossibility of navigating her in the tortuous river even while there was ample depth of water.

11th May: While I was away collecting in the morning Col. Racedo, two other officers and twenty-three men arrived from the *General Paz*. They brought their own provisions but none for us! With them came a man Aguirre who was to have joined the expedition at Timbó. He is said to be 'muy vaqueano' (very 'knowledgeable' or expert) and a fine hunter. The *Heave-up* is reported to be lying far down the river disabled with broken propeller shaft.

14th May: Dam V is now complete; a workman-like job—a backing of split palms, then bundles of grass and brushwood, and finally a facing of earth. At one side is an overflow which ought to prove a useful safeguard. These dams are useful to me as bridges for crossing the river, while the brushwood forms a trap in which the river fish occasionally become entangled.

19th May: Again at a full stop for want of water. In difficulties also from lack of drinking water, as well as indeed of provisions, which are now very low. Page has sent off a party of men down river with dispatches.

An unpleasant fight between Pool and young Page which resulted in disciplinary action against Pool—the old man threatening to brain him with a hatchet and expounding his view that 'the lives of you English are not worth a pin's head to me', and that 'several English have already had to be shot in this service'.

20th May: A very common bird is our old friend the Bien-te-veo, perched always in a conspicuous position. Another is the grey Tanager (*Saltator caerulescens*) with a short but not unmelodious song composed of two clear flutelike notes, followed by a chirpy trill and ending in two clear notes again, the last somewhat prolonged: *whew-whew-krr-ke-wee*. They are often seen pursuing one another and when doing so they utter a series of hard sharp chirps in rapid succession. Fresh Indian trails to-day and smoke in the distance.

23rd May: Held up by a broken connecting shaft which will take a day or two to repair so a start has been made on Dam VI. The food situation is serious and rations are being cut down to the minimum—an ounce of rice and four biscuits daily.

24th May: A paca-rail that I brought down had been hit by a single pellet which entered the head over the left eye. After a preliminary struggle it appeared to be overcome by sleep—a fish which it had caught remaining between its mandibles. It nodded its head, its lower eyelids raised, its head sunk into an easy position, its left foot slightly raised. It had all the appearance of being sound asleep: it occasionally 'smacked its lips' as a sleeping bird does. Occasionally it would open its eyes widely and then slowly raise the lower eyelids again.

When I hurried to get my camera the sudden movement caused it to wake up and fly off, but almost at once it came down, overcome with sleep. I duly preserved the head in spirit for laboratory investigations of the exact injury to the brain in this involuntary vivisection experiment.

25th May: This being the 75th anniversary of national independence the Commander appeared in the full glory of gold braided cap in place of his usual slouch hat: and the flag was ceremoniously hoisted in the morning and lowered at sunset. In further honour of the day our mess was presented with a bottle of pickles (mixed) and one of mustard (French).

The day being a holiday an expedition composed of Zorilla, Kennion, Jack, Nelson, César and Aguirre, started out after coffee to slaughter deer but unhappily didn't see any.

27th May: A strange new Woodpecker (*Leuconerpes candidus*) made its appearance: white plumage, with a patch of bare skin round the eye of a bright lemon yellow colour; the eye itself with white iris. Three were seen flying about and perching on palm trunks, uttering a shrill cry at short intervals.

Dissection showed the food to consist of honeycomb of a local wasp which collects honey. There was a remarkable tongue, not so far protrusible as in other woodpeckers, its terminal portion flattened dorsally and fringed on each side with brush-like processes, much as in the tongue of the Toco toucan but directed backwards.

28th May: The water level has fallen greatly and the *Bolivia* is high and dry. Not only is the new dam leaking but it looks as if something has happened to Dam V as the river is flowing with a rapid current. Nelson and César proceeded down river to see what had happened to the main body of troops who were left behind to put the dam in order and construct a stockade fort.

Rations are further reduced and our ship's company not in too good form: Aguirre, Agustin, Ochoa, Kennion, Diaz, all pretty ill.

In the afternoon Nelson and César returned with the startling news that the entire military body had cleared out with their canoes and stores, leaving behind a board nailed to a tree with the inscription Fortin Regimiento 6 de Caballería! So that was that.

May 30th: Cold and wet. Agustin reported having seen yesterday a herd of about 200 javalís i.e. the large Peccary (*Dicotyles labiatus*) reputed to be the most dangerous of South American mammals, fearsome stories being told of hunters surrounded by them and torn to shreds by their sharp tusks.

In the afternoon I went off in a canoe upstream and saw a couple of new blackheaded Parakeets (*Conurus nanday*) which came and perched in a tree top. I had only the rifle with me so had to get into such a position that the head end of one and the tail end of the other escaped destruction by the bullet. It is a beautiful species: less shy than most: utters shrill short cries.

On my return journey paddling silently down-stream I surprised a black cat (Eyra hu: *Felis jaguarundi*) and got within ten yards of it when it was startled by the canoe striking against some submerged obstacle and it bounded off into the vegetation before I had time to seize my rifle.

Arriving at the *Bolivia* I was greeted with the pleasing news that Venezuela, one of our soldiers, had disappeared along with his arms and an abundant supply of biscuits stolen from the small remnant of our store. No doubt if he reaches his fellow deserters down river he will be received with open arms. In the meantime we are ordered to shoot him at sight.

Apart from this desertion things do not look bright: we are within sight of the end of our provisions and the ship's company is hungry and disaffected and all very emaciated, with the fortunate exception of the Pages and Zorilla.

1st June: A year to-day since I left home. The morning was cold and misty but it soon cleared up. In the early morning we had the excitement of hearing distinct rifle fire in the far distance and concluded that either Venezuela or the main party of troops were having a fight with the Indians.* Later on we saw the smoke of Indian fires a couple of miles or so distant. And the Commander enjoined us to be extra vigilant and to be constantly armed. During the night anything moving on the bank is to be challenged and fired upon.

During the early part of June very little progress was made, the channel being more and more overgrown with tree trunks and the water becoming less and less. Disaffection among the ship's company increased, not made less by the general ill-health and the starvation rations. The Commander and his second-in-command, Zorilla, had violent differences of opinion, the latter declaring the position to be hopeless and urging a retreat down river; Page on the other hand, and in spite of illness, remaining determined to push on and reach Bolivia or die in the attempt (Plate V).

It was however decided to send Zorilla down river to expedite the sending of a relief party with provisions, and on 12th June Higginson the carpenter was set to work on the construction of an extra canoe. After a fortnight's work he provided a very serviceable craft and on 27th June the party started off down stream—Zorilla, Agustin, César, and the Doctor who enfeebled by fever and insufficient food realised that he was doomed unless he could get back to civilised conditions. With Zorilla I sent off letters to friends at home—our first messages to the outer world since entering the Pilcomayo. I was myself pretty ill and had to remain *hors de combat* that day, until roused by a commotion in the afternoon when I looked out and saw to my astonishment the Doctor on the bank, a most forlorn and abject spectacle: dripping wet and covered with mud; without coat, with only the remains of stockings on his feet; trembling and weak, almost *in extremis*.

Later on, when we had got him tidied up and fed and rested, the Doctor was able to tell us his miserable story. When they had come to a point where navigation was difficult Zorilla had ordered him to get into the water and help to push the canoe onwards. He did his best to help but when at last they reached open water he was told to cut across a neck of land and wait on the other side. When the canoe appeared, instead of being re-embarked he was roughly ordered to tramp along the bank. He

* Long afterwards we were to learn that the sounds came from a fight between the Indians and an Argentine expedition which had been sent up from the south to relieve us (p. 161).

struggled on but was soon left hopelessly behind. He rested for an hour in his exhausted condition and then as there was no sign of the canoe returning to pick him up he realised that his only chance of life was to make his way back to the *Bolivia.* Fortunately and surprisingly he succeeded in doing so.

During the remainder of June and early part of July the time passed very tediously for all except myself who had always the local natural history with its ever new and fascinating details to fall back upon. The food position got steadily worse. Of eatables there was only a little maggot-infested charqui and maize left. I of course did my best to help with what I was able to shoot but this was now practically confined to birds, the larger animals having wandered away from our immediate neighbourhood to places where fresh water was still available. The health position too was bad; the Commander distinctly going downhill: the Doctor getting more and more dropsical: Higginson the carpenter with severe renal and abdominal pain.

July 18th: The Commander who had been completely incapacitated for several days realised after his brave and prolonged fight that his only hope lay in reaching the outer world and getting skilled medical treatment so Higginson was set to work to patch up our one remaining canoe, in readiness for his departure down stream.

20th July: Page left in the canoe with the two negroes Jack and Smith. He refused to take the Doctor with him and he assured us that he would return again when his health was restored. However as we shook hands I wondered if we should ever meet again. Jack returned about an hour after their start and took away Aguirre as an additional man.

The departure of the Commander reduced our ship's company to nine: Henderson—engineer, Luigi Vignoli—doctor, Nelson Page—midshipman, Pool—steward, Roberson—cook, Higginson—carpenter, Diaz—corporal, Ochoa—trooper, and myself. Of these Dr Vignoli was already in a very weak state as a result of repeated malaria attacks combined with insufficient food and aggravated by the brutal treatment he had received on his attempted escape down river with Zorilla. He was an intelligent and well-informed man, a graduate and member of the staff of the University of Bologna, who had come originally to South America in order to study diphtheritic disease in Buenos Aires. He had married there shortly before accepting the appointment of surgeon to the Escuadrilla Bermejo y Pilcomayo.

22nd July: I had been out after birds for some little time when I heard the sound of whistles and shouts in the distance, so thinking that something must have happened I began to retrace my steps towards the boat. Presently Pool appeared and rather excitedly enquired if I was all right and I replied that I believed so. It appears that three shots of mine had been taken to be a signal for help and this fussiness had been brought about by the discovery of a visit from Indians during the night! Four axes, a plane, a large box of nails and some clothes had been left ashore by Higginson where he had been working at the canoe—and this morning all had vanished—while on the soft ground were the prints of feet which had obviously never worn boots.

We were now a small ship's company: two of its members quite old and useless for a scrap, so this visit of Indians to a spot only 60 yards from the boat and unsuspected in spite of our all-night vigilance emphasised the need of continued and still greater watchfulness. We had every reason to believe that we were in the territory of the warlike Toba Indians and the fact that they had never shown themselves openly did not inspire confidence in their friendly intentions.

The food position had been eased a little at this time, being supplemented by an occasional jacaré, cierbo (*Cariacus paludosus*, the large swamp deer), or Muscovy duck.

It being still winter we suffered a good deal from the cold at night, I myself particularly so as I was at this time taking hourly temperatures day and night. On 6th August the temperature at 6.30 a.m. was only 27·75° Fahr.; the vegetation everywhere coated with thick hoar frost and the pools of water in the swamp had a thin covering of ice. At the same time we were not without signs of approaching spring such as in particular the acacias laden with their beautiful pale cream flower balls round which many butterflies fluttered in the sunlight.

16th August: When out hunting I shot a Bandurria and had an unpleasant adventure when trying to obtain fire by igniting moistened gunpowder from a cartridge by using my small pocket-lens as a burning glass. When the powder did at last go off I had inadvertently got my head too close to it and had to sit for an hour or two completely blinded and reflecting on my probable fate. However I was at last able to open my eyes and found them undamaged, though eyelids were badly burned, eyelashes and eyebrows burnt off, and beard, moustache and hair thoroughly singed—in fact no more damage than was necessary to make the lesson impressive.

18*th August:* When out after birds in the morning encountered a large puma pacing slowly along like a great cat, so I put down my shot gun and went stealthily after him, getting to within 20 yards before he became aware of my presence. The first shot from my revolver seemed to wound him slightly and my next three all missed as he went off. I had got him into a sharp bend in the river and learned the lesson that a puma will never wet his feet if he can help it, for he doubled back and dashed past me close at hand rather than ford the river. After getting well past he stopped and looked round at me reflectively and then disappeared into the long grass.

The Doctor is very bad to-day, his pulse 120 and breathing very difficult: he is extremely emaciated.

20*th August:* To-day I got a fine specimen of the Aguará guazú (*Canis jubatus*) the solitary wolf of South America with dark brown fur, white breast and large ears. We had some of him for dinner. When cutting him up I found he had only a single functional kidney, greatly enlarged, the other (right) being represented by a small cyst with cartilage in its walls and limy concretions. The animal was quite healthy and in good condition—corresponding with what happens in the human being where the removal of one kidney becomes in time completely compensated by increase in the size of the remaining one.

25*th August:* Wind southerly, cloudy and bitterly cold. Towards evening an enormous column of smoke rose suddenly behind the large monte and about a mile off. The fire lit up the sky all night, making it almost light enough to read. Evidently a great Indian signal: answered by two other fires farther off.

The estero nearby is now drying up completely so we had to fall back on a well dug in the centre of it.

26*th August:* Still biting S. wind and uniformly clouded sky. Diaz killed a fine four-year old buck this afternoon and Nelson promulgated an order that all skins of animals killed were to be delivered to him intact. There was nearly trouble when the Corporal suggested that Col. Racedo his own commanding officer had first claim. Nelson at once sent for his sword but fortunately the corporal shut up and there was no further trouble beyond threats of what would happen if there were.

27th August: Still the everlasting South wind and bitter cold. Disaffection is becoming steadily worse and is now I think really dangerous.

A lecture from Nelson on the subject of discipline with the announcement that in future he would be always armed and would at once shoot on the slightest sign of insubordination. Diaz to whom the remark was specially addressed is in a particularly dangerous frame of mind and I have much reasoning with him to prevent his taking the first opportunity to avenge himself.

30th August: The Doctor is far through. To-day he called Diaz and gave him his watch as a memento and the corporal expressed his gratitude in a way strange to English ears, saying he would extract his bones, place them in a bag, and carry them to his widow. The Doctor gave me too a little memento—his pocket case of instruments, also a document relating how he had been treated and he begged me to see that his remains were not maltreated after death as he had been in life.

2nd September: In the evening witnessed a novel phenomenon—a procession of about 50 large bats (*Noctilio leporinus*) winging their way upstream with soft and easy flight and forming a pretty spectacle with their big transparent wings and their undersurface gleaming rosy-red in the rays of the evening twilight.

6th September: The Doctor very bad. His voice is weak and childlike and occasionally he maunders away to himself in his broken Spanish. He suffers great pain, crying out most of the night, while brutal messages come to him from the neighbouring cabin to 'shut up that row'.

8th September: This morning the poor Doctor worn out by his suffering and reduced almost to a skeleton died—the corporal and I sitting by him during his last hours. Occasionally he would revive a little and we would see his lips move as he tried to say Grazie or Addio but at last the end came. Higginson made a rough coffin and we carried him ashore and buried him by the edge of the monte. It was a cold bleak drizzle with no sound but the mournful *coo coo* of a dove in a tree close by.

9th September: Conditions very bad. Weather still dull and miserable, cold South-West wind and leaden sky. A good deal of rain and the upper deck leaking like a sieve: drip, drip throughout the night so that bed and clothes are all soaked through. Everyone feeling pretty low, most suffering from chronic diarrhoea. Food on starvation rations. And the general disaffection is stimulated by the tales about the menu of the Commander's cabin retaining both quantity and quality. Picturesque accounts of Nelson's meals do not help.

10th September: In the afternoon collected birds—amongst others a fine Blue Jay, the first I had seen for a long time. Up here it is a very shy bird, very wary, discrying one afar off and giving timely notice to its companions by its repeated cry—*ca, ca, ca.* It has a very characteristic and exceedingly neat flight. Sandpipers, snipe and such birds have become quite frequent during the recent showery weather and are looked upon by some as sure indications of the coming at last of the much longed-for Creciente or flood season.

11th September: A welcome surprise to-day in the form of a plentiful crop of mushrooms, especially over burnt areas and where the palms grow most thickly. They provided us with a really satisfying breakfast and dinner. Unfortunately this was their only appearance.

We had an interesting variation of the 'black crow' story. Corporal Diaz came in and told how an enormous tiger with feet as big as his two hands—had approached Higginson who rushed away and climbed a palm, leaving his rifle behind. He, the corporal, had witnessed the whole scene. However when Higginson arrived, looking much less excited than we expected, he explained that the enormous tiger was merely a small tiger-cat too small to fire at with his rifle.

13th September: A fracas in the afternoon when Nelson proceeded to discipline Higginson by striking him over the back with the flat of his sword. Higginson said 'Strike away', whereupon he was given an ugly gash on the head, cutting open his forehead and severing the right eyebrow. I did my best to stitch it up with the only means available a blunt needle and black silk thread, washing the wound with potassium permanganate solution and dressing it with cotton wadding and Friar's balsam. He will be disfigured for life.

16th September: Bird collecting in the morning. Away to the S.E. saw a magnificent flock of 'Wood Ibises' (*Tantalus loculator*) soaring high up in great circles. These large birds with their pure white plumage and black wings have a majestic appearance as they glide round one after the other in ever-changing circles.

To-day I saw the Scissor-tail (*Milvulus tyrannus*) for the first time up here. Another bird that has become common of late is the Churinche (*Pyrocephalus rubineus*) a little fly-catching tyrant with head and undersurface of a rich scarlet so bright and pure as to be almost dazzling. It may be seen perched on the top twig of a bush or low tree, sallying out to catch a passing insect and then flying back to its perch.

Higginson's gash is progressing favourably but he is moody and downhearted, and no wonder, for he was a good-looking chap and rather proud of his appearance—now and for ever spoilt by the big cut through his right eyebrow and forehead.

17th September: During the night and morning the North wind blew with its usual effect of making one feel tired and fagged with dull aches all over one's body. However in due course it changed round through West to South-West and brought a great drop in temperature from 90° F. to 66° F.—a fall of 24° in one hour.

18th September: My birthday. I had been doing my laundry work down stream, and returning to the boat laden with my dried clothes and a bucket of water I was hailed by Nelson who was aloft at the Maxim gun and told to get on board at once as there were Indians around. It appeared that two Indians had been seen near the boat and had then hied off apparently to fetch their companions. I had no sooner got on board than the Indians reappeared, twelve in number this time, threading their way through the brushwood to the N.W. of the boat [Plate VI]. Although fully armed they seemed to be friendly and Nelson and I went ashore and a useful commencement of conversation was made by the Indians repeating over and over again the words Amigo and Tabaco. One whom we learned afterwards was a powerful cacique named Yordaik could also speak fairly fluently in the Guaraní language of Paraguay. He assured us that both he and another cacique who was with him simply loved the Cristianos, in great contrast to their neighbouring cacique upstream who was 'muy bravo'.

Apart from Yordaik the Indians were all very tall—several of them fully six feet. Their clothing a simple strip of coarsely woven cloth round the loins except that two had in addition a kind of vest of open cord work. Round their waists they carried a small bag of the same material holding apparently odds and ends. Their skin where exposed was dark red; the face painted with bright red pigment. Their feet were bare: their heads covered with thick and luxurious and obviously well-kept locks curling over their neck behind and in the form of a fringe over the forehead. Four of them had hats formed out of strips of palm leaf and the tall cacique was greatly gratified by being presented with the doctor's 'bowler' and white shirt which he at once donned to the great admiration of his followers. For arms they all carried bow and arrows while one had a formidable looking tomahawk formed of a cold chisel stuck in a club.

The Indians left us just before dark and encamped at the edge of the monte near the doctor's grave and we could see the glimmer of their fire throughout the bitterly cold night.

They stayed with us three days during which we had very friendly intercourse. They bartered some ostrich feathers and deer skins and, better still, some food—about thirty Guinea pigs the product of their first afternoon's hunting, and the flesh of three cierbos.

One interesting fact that emerged from our talks was that the Indians had been keeping careful and constant watch upon all our movements. While away hunting I had all the time been followed and watched though the watcher with his dark coppery skin and stealthy movements had remained completely unsuspected.

On the forenoon of the 20th the Indians left us with expressions of friendship and promising to return next moon with their 'little brothers'. We felt however no particular enthusiasm for seeing them and their brothers as we could not help suspecting that our visitors had been sent mainly to spy out the land and that when they came again it would be to attack and annihilate our little party.

The period following this visit from the Indians was marked by ever increasing weakness, depression, and disaffection among our company.

1st October: About mid-day a dark cloud resembling an approaching snowstorm appeared in the South-West. It rapidly came up and turned out to be a swarm of locusts, or migratory grasshoppers as an entomologist would prefer to call them. In a short time they were right over us—a dense cloud, obscuring the sunlight and producing a loud rushing noise as of an approaching tempest. At a little distance the insects resembled the falling flakes of a heavy snowstorm, while when close by their wings flashed and glittered in the sunlight. For over an hour and a half there was no apparent diminution in the multitude but after that they became sparser though continuing to pass until about 4 p.m. The latter part of the army were possibly individuals which had paused to rest and feed and deposit their eggs before hurrying after their companions. Trees on the line of advance were literally covered with the insects all feeding voraciously on the foliage, while green patches of herbaceous vegetation were coloured brown in the same way.

4th October: Saved!

Pool and I went off at dawn in search of game but were unsuccessful and were returning to the ship tired and weary and nearly devoured by mosquitoes, when suddenly we heard close at hand an extraordinary sound which made us look at one another in amazement and then burst out laughing as we realised what it was—the opening notes of a cavalry trumpet call—and that it meant our rescue from what had seemed certain death, by starvation or by the Indians. I at once let blaze with both barrels of my express; answered immediately from the ship. Then Pool dashed off frantically towards the *Bolivia* while I followed more slowly, drinking in the delicious jingling of mule trappings approaching at a sharp trot. I had just reached the *Bolivia* when our rescuers appeared coming along the narrow pathway—a lot of strange faces with the familiar countenances of Jack and Aguirre in front. Besides the latter there were an officer (Lieut. Candioti) and twenty men in the blue uniform and red caps of the Argentine cavalry, each man with a bundle of fresh beef strapped to his saddle. After the exchange of heartfelt congratulations and thanks they told us the news.

In the first place Captain Page had died on the 8th August on the way down the river, nineteen days after leaving the *Bolivia.* He had been getting worse each day; his legs very soon becoming paralysed. Arrived at the point where the *General Paz* had been left, there were no signs of her, and this disappointment no doubt hastened his end.

On 8th August they stopped earlier than usual and Page asked them to cut a lot of grass to make a good bed, for he felt weary and tired and thought he would sleep sound and long that night. For the last few days he had had much fever in his head and the upper parts of his body, while the lower parts were deathly cold. They lifted him out of the canoe and put him on the bed and Smith bathed his forehead with cold water. He had been groaning much in the extremity of his suffering but gradually quieted down and seemed to sleep. Noticing an unusual stillness Smith got a light and found the old man dead: his expectation of sleeping long and sound that night had come true.

The three men, fearful of the fate awaiting them if they reached the mouth of the river and merely reported the loss of their chief, saw that they must try at all hazards to get his body down to Puerto Pilcomayo where it could be examined and death shown to be due to natural causes. So by 8 o'clock that evening they were again under way with the body on board, and it was not until after almost ceaseless paddling through four nights and three days that they reached Puerto Pilcomayo—the body in an advanced stage of decomposition so that the canoe with its contents had to be buried at once.

Listening to the recital of that grim story of the canoe journey with the decomposing corpse on board and of Page's sufferings before the end came, I for one tended to be forgetful over the less worthy traits of our former leader and to remember rather his better side. His hard relentless nature was only one aspect of that inflexibility of purpose which together with his personal bravery and his skill in the technique of river navigation made him a really great explorer.

Our rescuers brought news from the outer world. Zorilla who had been sent down to expedite relief to us had duly reached the mouth of the river but finding that the great 1890 revolution was in progress and that he was just in time to take command of the last government vessel going down river, proceeded to do so, regardless of those left behind on the Pilcomayo. He had reported, we were told, that the doctor had gone mad on the journey down the Pilcomayo, had jumped ashore and been attacked and devoured by a tiger! He had also reported that he had left the remains of the expedition dying of scurvy.

We were told that an expedition had been sent out to relieve us several months earlier with supplies of provisions: that they had been attacked by Indians, that they had sent down their wounded but that now for three months nothing had been heard of the remainder and it was feared they had all been killed.

That the need of a second relief expedition had been urged when Zorilla's report had become known. The Argentine Minister in Asunción had used his influence, backed up by pressure from H.B.M. Minister in Buenos Aires, and finally the Paraguayan Government had decided that it would itself send an expedition should the Argentine not do so. However Col. Uriburu, Governor of the Chaco, took matters in hand and dispatched a new relief expedition; an officer and 20 men of the 12th Cavalry with 40 mules and 10 bullocks—Jack and Aguirre with them to act as guides. It was regarded as a forlorn hope as the *Bolivia* was known to be in a territory of Indians of the most treacherous reputation and the expedition had practically been given up for lost.

However the relief party has after all succeeded in reaching us, after a toilsome journey of over a month owing to going astray in its earlier stages, reaching the tortuous Rio Confuso, wandering backwards and forwards along its banks, and finally marching back for seven days before again finding the Pilcomayo.

Our first impression of the new arrivals has been their remarkable plumpness—even obesity—our standards of such things having no doubt changed considerably during our period of semi-starvation. That was now at an end. The bullocks were enormous creatures in magnificent condition, but our first meal off fat fresh beef, though providing us with a great memory of its delicious first taste, proved almost too much for our enfeebled digestive arrangements and we ate but sparingly.

5th October: Nelson busy all forenoon with preparations for his immediate departure. He takes with him seven of the relief troops and also Jack and Aguirre. He goes to Buenos Aires to report and says that he hopes to return to us. 'I will not abandon you' were almost his last words. Henderson the engineer was left in command, with orders to do his best to get the *Bolivia* down river and in the event of this proving impossible, to burn the vessel and contents. With all my collections on board this would indeed be a disaster.

About 2.30 the party left—Nelson, a corporal and six men of the relief party; Ochoa one of our original lot of soldiers, Aguirre and Jack. They were, of course, travelling overland and took with them 13 mules.

The departure of Nelson Page changed the whole psychological atmosphere on board. 'A hungry man is an angry man', and the disaffection on board the *Bolivia* had become fanned into a dangerous flame by the report that the three superior officers of the expedition, instead of sharing alike the rigours of semi-starvation, were being well-fed from a private store of provisions. It was no doubt felt essential that they should keep themselves fit but this feeling did not appeal to the rest of the ship's company. It transpired that Kennion had actually made representations to the Commander who bluntly told him that he would shoot the first person who again ventured to make any complaints. Knowing as I did the violence of feeling on board I was in constant dread of the dark threats uttered, particularly against the younger Page, being carried into effect.

One of Henderson's first actions was to investigate the cache of provisions stored under the floor of the Captain's cabin and transfer its contents to the general store. One of the most welcome changes was the renewal of our acquaintance with tea and coffee!

Apart from matters of food, living conditions were pleasantly changed. The duty of watch keeping was taken over by the soldiers and it was pleasant drowsily to hear at 15-minute intervals through the night the signals of the outer circle of sentries—a short blast on a whistle made out of an empty cartridge case; and when morning came, to feel really rested now that the main strain of responsibility had been taken over by others.

The new lot of soldiers looked tough and efficient. Their unit, the 12th Regiment of Cavalry, was regarded as one of the best of the Chaco regiments and their commanding officer Lt. Candioti came also with a good reputation. A bit of a martinet, his men seemed patterns of discipline—in contrast with the sloppiness of our previous lot of soldiers.

The availability of the mules brought by the relief expedition facilitated my getting about over longer distances than had been possible on foot. They were very hardy as compared with horses in their resistance to the Chaco climate. They were very wild and would often fight hard to resist being mounted. In bad cases we would have to draw the mule close up to a palm with the lasso wound round it. Having been successfully saddled and mounted the mule was walked round and round the palm so as to unwind the lasso and then when at last free it would dash off bucking and doing its best to unseat its rider. Though not an expert rider I developed quite a liking for the mule. A certain femininity suggested that a line of *Rigoletto* might we modified to read 'la mula è mobile'. At one moment she would be all friendliness, another would do her best to kill one. One time I was crossing the southern branch of the Pilcomayo carrying on my saddle photographic apparatus, a saucepan and other equipment rather as the White Knight in *Alice through the Looking Glass*. We slithered down the steep bank of the river, forded the stream and proceeded to clamber up the far bank, but just as we were reaching the summit the girth broke and I and all my paraphernalia rolled down into the stream. I seemed to have a distinct suspicion of a flicker of humour in the eye of that mule as she looked back and watched my descent into the stream.

About a week after the arrival of the relief party we became aware of Indians in proximity but they did not attack nor did they show themselves openly until just before sundown on the 13th when one of the soldiers by the camp fire suddenly exclaimed 'Ahi vienen los paysanos'—there come the natives. Everybody sprang up and as they approached nearer we recognised the faces of some of those who had visited us 'last moon', and we were soon in friendly converse with them although the fact of their being fully armed and without their women folk made us suspicious as to what their original intentions had been.

From now onwards we had Indians with us almost continuously during the hours of daylight. They watched all our doings and there was much bartering.

I took the opportunity of taking photographic snap-shots which operation at first did not cause the Indians any alarm: the camera was no more strange to them than other parts of our equipment. Gradually however it began to excite their special interest and suspicion, so when photographing a particular individual I would take care to keep my attention apparently concentrated on some one else. Plate VII (*a*) shows one of the Indians sitting quite at ease and showing merely a mild interest in what might happen to his brother but with no idea that he was personally concerned. With further experience fear and suspicion of the camera increased and this is well illustrated by the two consecutive portraits of Chinkalrdyé (Plate XV (*a*) and (*b*)) the second showing the change in expression before he started up and brought the sitting to an end.

I also busied myself with acquiring a speaking knowledge of the Indians' language, commencing by noting down phonetically the word uttered on my indicating some particular object. In this way I soon accumulated a considerable vocabulary of nouns —to be followed later by the more evasive ideas expressed by adjectives and verbs.

The Indians very soon began to show a distinct difference in their indications of friendship towards myself as compared with my companions, and I was inclined to attribute this to the fellow feeling that the hunting of wild animals was my chief interest as it was theirs.

A great event was my first hunting expedition alone with the Indians. Yordaik had at various times been telling me how his men wanted me to go off on a hunting expedition with them. At first I was doubtful in view of the Tobas' reputation for treachery but reflecting how completely I had been in their power on my innumerable lonely collecting expeditions I decided to risk it.

So we started off one morning to hunt ostrich or rhea—'mañik' in their language, 'ñandú' in Guaraní. I carried my rifle, and I gave my shot-gun to Chimaki—my special friend—to carry, and with us came Chinerutaloi and Yaraitlik. Chimaki led the way along a pathway to the north-west, going along at a swinging pace, as the Indians do, nearly 5 miles an hour, walking with elastic step, body erect and head thrown back. After about 3 miles we left the pathway and struck across open grassy camp and at times skirting the edge of the monte. After a bit Chimaki pointing to a bit of country with an occasional tree or bush and many ant-hills said this was the country of the mañik. We now separated into two groups Chimaki and I going off to the right. We had gone but a short distance when Chimaki peering intently to the left indicated that there was a mañik. I looked but perceived no mañik: only the grass and an occasional grey ant-hill. We had not stopped and now Chimaki started coursing round in a circle of 30 to 40 yards radius, making all the while excited gestures towards the centre of the circle. Then suddenly things happened. I became conscious that one of the apparent ant-hills was surmounted by the slender neck and head of a ñandú which rose to its full height and just as it dashed off received one of my express bullets in its sacral region, being killed thereby instantly. Chimaki gave a triumphant shout and in a few moments we were joined by Chinerutaloi and Yaraitlik.

The bird was a large male and close by we found the nest, built of earth and grass mixed, raised about 6 inches above the ground and containing thirty eggs—the male ostrich taking over the incubating duties from his numerous harem. The eggs I looked on as a magnificent haul but my joy was lessened when on breaking one open we found a well-developed ostrich chick inside. However about two-thirds of the yolk was still there so the Indians quickly broke the tops of several of the eggs, fished out the little ñandús, and stood the eggs to cook round the fire which Chinerutaloi had lighted with the aid of his fire-sticks. Part of one leg of the ostrich was also put to roast and during the process of cooking Chimaki busied himself pulling out and collecting the feathers while Chinerutaloi cleared out the entrails. I meanwhile lay on the ground watching operations and enjoying myself immensely. The meal was soon cooked and we fell to 'con mucho gusto'. Then an hour's rest and I gave the word 'Kolŭk'—let us go. The Indians packed the remaining eggs and bits of the ostrich into network bags and, heavily laden, we made our way back to the ship and were welcomed with much acclaim.

During the weeks following the arrival of the relief party the *Bolivia* remained at Fort Page. Indians came and went, watching our doings, bringing goods for barter and very soon having sufficient confidence to lay aside their arms and even to let their women and children come round the ship. Most of my own time I spent hunting with the Indians, adding to my collections, and studying the local natural history of which more will be said in the next chapter. Close association day after day with the Indians led gradually to the development of deep friendship. During my hunting expeditions with the Indians it would every now and then happen that one of them would recall how so many moons ago I had killed such and such an animal at that particular spot, and I was made to realise how my supposedly solitary collecting expeditions had been in fact carried out under constant surveillance and there could be no doubt that the reports by my unseen watchers as to my interest being concentrated not on looking for signs of their own presence in the neighbourhood but rather upon animals and plants had much to do with inspiring that fellow feeling which as I believe formed a sound foundation for the mutual trust into which our relations developed.

25th October: The commission arrived back from Puerto Pilcomayo. They reported that Nelson had gone off to Buenos Aires to make his report and receive orders, but that there was no news from him although they had waited ten days. The journey from Puerto Pilcomayo to Fort Page (Dam VII) had taken five days.

29th October: This morning a new commission set out for Puerto Pilcomayo—Sergeant Cuña, six men, and Pool, so that Henderson, Kennion, and myself are the only Britons left with the *Bolivia*.

In the evening one of the older Indians performed one of the long monotonous chants intended to keep away the powers of evil, the monotonous effect being accentuated by his rhythmic beating time with a small gourd with dry seeds shaking about in its interior. At regular intervals the chant rose into long high-pitched refrain shouted with the entire strength of the performer's lungs. When heard from a distance it is only these wild wail-like outbursts that are heard, sounding doubly weird in the midst of the deep silence, their only accompaniments being the croaking of the frogs and the deep *boo-hoo-hoo* of the Ñacurutú.

3rd November: Candioti had announced yesterday his intention of paying a visit to-day to the camp or tolderia of Cacique Yordaik. However this morning we found that the Indians had all cleared out during the night—no doubt to warn their friends and maybe prepare a warm reception for us. Nevertheless the Lieutenant decided to adhere to his original plan and about 6.30 we started off with six soldiers including two corporals regarded as 'muy vaqueano'—very knowledgeable—in the art of following trails. Each of the men carried 200 rounds of ammunition and we also took a sufficiency of charqui to keep us going for a few days. We started off following the fresh trail made by the Indians last night but before long it disappeared the Indians having scattered in order to delay any pursuit. However there was an older well-marked path leading in the same direction to the South Westward so we followed this. Our way lay through varied scenery—now through a wide open and treeless expanse dotted by ant-hills, now skirting along the edge of the monte duro, or again through parklike scenes dotted with trees and bushes, or anon through the ordinary palmar. Nothing special happened beyond alarming a small Puma and causing him to spring up from his slumbers in the grass and clear out as we approached.

At a distance of 5–6 leagues from the *Bolivia* we reached the much-talked-of Southern branch of the river, lined on each side by esteros with tall reeds and with steep banks some 20 feet in height clothed with a tangled growth of low trees. Our first view was of a beautiful reach of six or seven hundred yards in length reminding us of the lower parts of the Pilcomayo below Las Juntas. We were able to ford the stream without difficulty as the bottom was fairly firm, the water, clear and salt, reaching well up over our saddle flaps. The estero on the far side was more difficult as it still had a good deal of water in it and the leading mules got into difficulties sinking down into the mud so we had to retrace our steps. However we at last discovered a kind of pathway obviously used by the Indians which led through the estero, and by this we reached the far side. It was now midday and fresh water being available we called a halt, unsaddled and tied out the mules to graze, while one of the men prepared an excellent breakfast a guizado or stew of charqui and rice, followed by roast Cierbo. After a couple of hours siesta we saddled up again and pursued our way westward until we suddenly perceived an Indian at the margin of a piece of monte. We at once closed up and signalled to the Indian to approach which he did and having got a friendly reception was joined by about 15 companions. They were soon all shaking hands with us and protesting their friendship—but I noticed that many of their hands trembled a good deal showing that after all these men have 'nerves' like those of other races. Scattered about in the grass were a lot of cierbo hides and other things which they had hurriedly cast away when we had become visible in the distance. Our meeting having been so friendly the women of the party also emerged from the monte and it was explained that they had been actually on their way to the *Bolivia*. This being so we decided that we had better return to the boat and leave our projected call upon Yordaik for another time. Our turning point we estimated to be about seven leagues from the *Bolivia* which we duly reached without incident about six o'clock after a good long day—about ten hours—in the saddle.

7th November: I have at present four pets given me by the Indians. Kūss is a young white-lipped Peccary which follows me about like a dog, keeping up a conversation of friendly grunts the while. Toba is a young Nutria, a most friendly little creature which at meals sits on its hindquarters gnawing at a piece of food held in its forepaws. The other two are birds, a young eagle—Buzzy—which inhabits a coop along with an enormous young ñacurutú owl. Buzzy has a terrific appetite and whenever he sees me in the distance returning from a hunting expedition begins to shriek out food, food! Unfortunately he apparently made the discovery that he could secure my attention by bullying the unfortunate owl, pulling out tufts of his down while he blinked his great eyes, not knowing quite what was happening and merely remonstrating by a feeble hissing sound.

11th November: Today two new animals not hitherto seen—a crab-eating Racoon (*Procyon cancrivorus*) and a magnificent Cuckoo (*Crotophaga major*), an enlarged edition of the Common Ani, of dark glossy purplish steel blue with concentric rings on its feathers of brassy green—altogether in appearance perhaps the finest bird I have so far collected.

22nd *November:* In the evening the Commission with Nelson Page arrived from Puerto Pilcomayo, bringing with them a large mail and half a dozen bullocks. Nelson marked his resumption of command by immediately placing Mr Henderson under arrest with orders to furnish a detailed report as to his administration while he was in command. The trouble reached its climax when an enquiry was held next day from which it appeared that Kennion had accused Henderson of having plotted to make a prisoner of Lieut. Candioti on his showing any intention of abandoning the ship. I was suspected of being in league with Henderson in his nefarious design. It was a flattering tribute to the aged Henderson and myself that we should be thought capable of taking Candioti and his fifteen soldiers into custody but after a little investigation it was decided that the whole thing was a silly invention, possibly inspired by the idea of settling an old grudge. On the following day Kennion left us for good, accompanying a commission of Corporal Gomez and four men taking dispatches down to Puerto Pilcomayo.

1st December marked the start of the *Bolivia* downstream from the never-to-be-forgotten Fortin Page (Dam VII). It is therefore appropriate to break off my diary and say something about the natural history of that region and its Indian inhabitants.

CHAPTER VI

FORTIN PAGE

To obtain a fair picture of the natural history of the interior of the Gran Chaco one cannot in fact do better than examine the region round Fortin Page, the farthest point reached by the *Bolivia*. Its most characteristic feature is the palmar (Plate IX (*a*)), a practically dead level plain covered with coarse grass 1–3 feet high and studded with Fan Palms (*Copernicia cerifera*; Carandai of the Paraguayans, Palma negra of the Argentines, Chaik of the Tobas) averaging about 30 feet in height, the trunk about 7 inches in diameter. In the young palm the spiny bases of the leaves remain in position round the stem and serve to protect the soft succulent heart from the attack of deer and other vegetable feeders. No doubt the wind pressure on the leafy head is an important limiting factor to the palm's increase in height, but although the average limit is about 30 feet there is a considerable amount of variation. The two tallest specimens that I encountered measured 72 feet 6 inches and 62 feet 10 inches respectively. Occasionally one sees branched specimens. One such bifurcated at about 45 feet from the ground, one of the two branches bearing a normal leafy crown while the other bifurcated twice and finally split up into about twenty terminal branches. The stem of the Carandai is dark in colour and very hard. It formed our chief building material when constructing stockade forts or huts.

Over wide areas the palm stems are seen to have a dark stain marking the level reached by flood water, an indication of the fact that this species of palm favours ground subject to occasional flooding. It does not occur, as is the case with the hardwood forest, on ground above the flood level nor on the other hand does it occur on ground where submersion in water is the normal condition.

The scenery of the palmar when first viewed produces a powerful impression, a certain weirdness and loneliness peculiarly its own, and accentuated by the mournful rustle of the dry palm leaves shaken by the wind. It is seen at its best in the early morning when the atmosphere is intensely clear and transparent, when there is not a breath of wind and the rays of the rising sun illumine the palm tops with a rich golden light. Then indeed the palm forest assumes an ethereal and almost fairy-like beauty. On the other hand a dark and gloomy winter's day when the sky is of lead, when the grass is withered or burnt by Indian fires, when the cold and biting south-west wind hurries across the plain—knocking the palm-leaves together and emphasising their dismal rustle—then the picture is bleak and depressing.

Conspicuous in the palmar are scattered solitary specimens of a large dicotyledonous tree, the Vinal (*Prosopis ruscifolia*), much branched, wide spreading, gnarled and knotty, with rough bark deeply furrowed vertically and hard reddish heart wood. A good specimen measured 50 feet in height and 10 feet in girth at 4 feet from the ground.

The most interesting feature of the vinal is its enormous straight, pointed spines. A tiny twig only an eighth of an inch thick may bear a spine half an inch thick at the base and 5 inches in length. To those botanists who regard spines as starved branches these spines of the vinal, branching off from twigs much smaller than themselves, must constitute a troublesome problem.

It may well be that the enormous spines of the vinal are a relic from the days when this region of the world was inhabited by gigantic herbivorous Edentates against which they served a defensive function. It was interesting to note that they are now put to a different use, many of them being hollowed out and converted into a safe abode for small colonies of ants.

The palmar possesses its characteristic fauna. Of mammals the Pampa Deer (*Cariacus campestris*), the Puma (*Felis concolor*), and the little wild Guinea-Pig (*Cavia aperea*) are the most conspicuous. I noticed an interesting camouflage habit in the last-mentioned. Ordinarily when alarmed it trusted to its great rapidity of movement but more than once when it became aware of my presence in its neighbourhood it proceeded to glide into concealment with movement so slow as to be hardly perceptible. The stoppage of all movement is a common device in the animal kingdom to escape notice. In many of the smaller animals it is often misleadingly spoken of as ' feigning death '. The mere slowing down of movement is less familiar except in the case of animals approaching their prey. In the case of the large cats the camouflage afforded by this slow movement is I have no doubt helped by the distractive effect of the movements of the tail tip—far removed from the ' business end ' of the creature. To the hunter or the naturalist the avoidance of sudden movement is of course of the greatest importance.

Less conspicuous but of greater scientific interest are some other of the small mammals. The most fascinating of these is a little Opossum (*Marmosa pusilla*) no bigger than a large mouse, with big protruding black eyes in correlation with its nocturnal habits. It makes its nest in the hollow of a prostrate palm stem and its reproductive arrangements are intermediate between those of the most primitive mammals (Monotremata) which still lay eggs and whose young go through the greater part of their development outside the body of the mother, and the ordinary modern mammals in which the egg is retained within the mother's body for a prolonged period and the young brought forth when they have already reached an advanced stage of development. In this little mouse opossum the young are indeed ' born ' but at an extremely early stage of their development. Each one becomes attached to one of the mother's eight teats which are arranged in a circle or rather an ellipse on her under surface and hangs on to it while it proceeds with its development. During this period the young are not sheltered in a pouch as they are in typical marsupial mammals such as the kangaroo: the pouch has in fact not yet become evolved in creatures such as Marmosa which represents an intermediate stage between the egg-laying and the pouch-bearing mammals. After completing its development the young Marmosa ceases to hang continuously on the teat but it does not free itself completely from the mother. It remains holding on to her, the end of its prehensile tail firmly curled round her tail, and its highly movable fingers and toes grasping her fur. The young remain so attached to the mother until nearly full grown, and a Marmosa family of half-a-dozen or more individuals—the progeny hanging firmly on to the mother and hardly distinguishable from her in size—presents a remarkable spectacle.

Marmosa is not the only marsupial common in the palmar and woods. Two others occur, the Comadreja already mentioned as occurring on the Pampa, the largest of all the opossums, and an intermediate species (*Didelphis crassicauda*). Of these the first mentioned has a completely formed pouch in which the young remain attached to the teats for a prolonged period, while the second possesses not a complete pouch but the beginnings of one, in the form of a couple of folds of the skin, one on each side of the under surface of the body.

Among the Chaco mammals of special scientific interest are a few survivors of the group Edentata which provided in past times the gigantic Megatheriums, Mylodons and Glyptodons. Of these the strangest looking is the great Ant-eater (*Myrmecophaga jubata*) with its extraordinary head prolonged into a curved snout with the tiny mouth opening at its tip, and its long tail clad with great coarse hairs so long as to give it an apparent thickness equal to that of the trunk.

Much of the ant-eater's time is spent in slumber among the long grass, lying on one side with tail folded forwards so as to cover the head. When active he proceeds at a sluggish kind of trot with the tail projecting horizontally backwards in line with the body. The fore limbs are enormously powerful, armed with huge claws, and turned inwards so that only the outer edge of the foot touches the ground. They are used to embrace and tear open the ant-hills whose inhabitants, especially termites, constitute the animal's normal food. The ant-hill being torn open the ant-eater nuzzles about among the crowds of termites and ants, pushing in among them his slender tongue which can be extended over a foot beyond the mouth and is covered with sticky saliva secreted by enormous salivary glands which extend far back under the skin of the chest. The tongue becomes promptly covered with a struggling mass of insects which are drawn into the mouth and disposed of.

The young Myrmecophaga is carried about for a prolonged period on its mother's back but I was never lucky enough to see this picturesque spectacle.

The flesh is eaten by the Indians and the skin is valued on account of the toughness and impenetrability which make it specially suitable as a protective of the legs when traversing spiny undergrowth or sharply cutting grass.

Two species of bats attracted special attention on the Pilcomayo. One a large *Noctilio* (*N. leporinus*) was to be seen at dusk hurrying up or downstream in flocks of thirty to fifty towards their feeding area, or later flitting about with noiseless flight just above the surface of the water, every now and then dipping down to the surface and rising again with a captured small fish—surely a remarkable habit to have developed in a flying mammal.

Again our mules quite commonly showed by streaks of clotted blood down their withers that they had suffered during the night from visitation by Vampire bats (*Desmodus rufus*). However, though I kept watch for them night after night I never caught a glimpse of one: and none of our company, even when sleeping without mosquitero, was ever attacked.

Of armadillos the three met with in the Pampa were again found in the Chaco. The Indians hunt them by simply following up their trails and clubbing them. The Indians had tales of the larger Armadillo (*Priodon gigas*) occurring occasionally in their territory but I never encountered one.

Of the characteristic birds of the open palmar there come next after the Ostrich (*Rhea americana*) the Martineta Tinamu (*Rhynchotus rufescens*) and the Chuña or Seriema (*Cariama cristata*) the voice in each case providing one of the characteristic sounds of the palmar. The plaintive whistle of the martineta showed a slight but quite recognisable local difference from that of the same species on the pampa. The voice of the chuña is a harsh wild cry.

During my stay six years later among the Mushcui Indians they brought me a young chuña which grew up into a charming pet. When he accompanied me on my walks abroad I would carry with me a slender spear to transfix frogs which formed his favourite food. It was interesting to see that the ones he preferred were those equipped by nature with obliterative colouring and active movements. When offered the slow-moving *Atelopus* with its bright warning coloration of black, yellow and red he would regard it for a moment though with obvious disfavour and would have nothing further to do with it.

During siesta Tum-um-hit would sit on my pillow, keeping up all the while a continuous friendly cheeping. At night he roosted on a beam just above the foot of my bed. It was amusing to see his reluctance to get up in the morning. On seeing me active he would rise on his perch with a determined expression but after a moment would flop down again and shut his eyes firmly. Then he would repeat the performance, perhaps more than once, but at last he successfully roused himself, flew down from his perch and then treated himself to a good chase after the missionaries' fowls, which fled in all directions, scared by his hawk-like appearance.

Poor Tum-um-hit came to a sad end. One morning he was found lying dead by the side of the pool in front of the hut. One of the Indians related how he had seen him come down from his perch during the night and drink deep draughts of water; and the post-mortem disclosed in his gizzard the skin of a rat which had been preserved with arsenical soap. Tum-um-hit had in fact been the bane of my colleague Budgett's life, for as he sat at his laboratory table, making dissections or preparing some rare specimen, Tum-um-hit would keep walking round at a distance, watchfully regarding him the while. Then when Budgett, having turned away to consult a book, turned back again to his dissection, he would find it gone, snapped up by Tum-um-hit whom he could see disappearing in the distance.

One of the most characteristic sounds of the palmar was the loud roll as of a kettle-drum, produced by Boie's Woodpecker (*Campephilus leucopogon*) common alike in palmar and monte and conspicuous from its large size, its black plumage with deep creamy buff tippet down the middle of its back, pale yellow iris, and above all the magnificent vivid scarlet of its head with large pointed cowl. It delights particularly in making use of the hollow stem of a dead palm which acts as a resonator as it drums on it with rapid blows of its heavy beak. Its ordinary voice is a low soft and squeaky chattering, uttered when there are two or more in company. (See also Chapter v, p. 59.)

Other woodpeckers than Boie's were conspicuous too in the local bird life. *Dryocopus lineatus*, also large and black with scarlet cowl but without the buff tippet, had a pure white band running down each side of the neck. *Caleus kerri* also of large size but with shabby-looking plumage of faded brown and rusty hues, a soiled and bedraggled-looking cowl of pale yellow, dark red iris and a strong musky smell.

Still other species showed a great variety of brilliant plumage; *Chrysoptilus cristatus* with lemon yellow plumage thickly dotted with round black spots; *Chloronerpes chrysochlorus* deep golden olive with scarlet cap; *Leuconerpes candidus* pure white with deep black wings and upper back, a patch of bare skin of bright lemon-yellow round the eye, and a white iris—a feeder on honey.

A specially interesting woodpecker was the ' Carpintero ' of the Pampa (*Colaptes agricola*) which here, as there, has emancipated itself almost entirely from the tree-frequenting habits of the ordinary woodpeckers. It was occasionally seen on trees, sometimes clinging vertically to the bark propped up by its stiff tail feathers, in normal woodpecker fashion, but at other times—and more frequently—perching crosswise after the manner of ordinary perching birds. More usually however it was to be seen searching for food on the ground, running about in a manner impossible to the typical woodpecker in which the legs are shorter than in this species. Most frequently it was in small flocks—sometimes of as many as a dozen. It uttered a clear piping cry—*whew-ew-ew-ew*, and dug up its insect food from the ground or from the ant hills. In most of the specimens I examined the food had been of black ants common in the usual ant-hills of the palmar.

Curiously, that other large Woodpecker *Chrysoptilus cristatus* showed signs of diverging similarly from the normal woodpecker habits. It was observed at many points on the Pilcomayo and in early spring was quite numerous, frequenting especially the open woodland and feeding sometimes in trees—perching indifferently woodpeckerwise or not—but often on the ground. In the early spring, about the beginning of September, many were to be seen inspecting hollow palms with a view to nesting.

Last of the woodpeckers to be mentioned is the tiny *Picumnus pilcomayensis*—a new discovery. Like other woodpeckers it makes itself heard by rapid drumming with its beak, but the faint whirr so produced is very different from the loud roll of its comparatively gigantic relative Boie's woodpecker.

A delightful element in the Chaco bird fauna was contributed by the parrots and parakeets of which six species were conspicuous in the neighbourhood of Fortin Page. The little parakeet *Conurus acuticaudatus* was abundant especially in spring and summer—going about in pairs, several in proximity forming a small flock, and attracting attention from a distance by their characteristic voice.

The Black-headed Parakeet (*C. nanday*) was at times the most abundant, usually in flocks of as many as fifty or sixty, and making the whole air resound with their shrill sharp screams. They were often to be seen in association with the next species (*Bolborhynchus monachus*), among the bushes or on the ground close by, or on the other hand feeding on the berries of the parasitic Loranthaceae (mistletoe) common on many of the local trees. Both of these species are less shy than the others and when one of a flock was wounded its companions would gather round with shrill cries of protest.

The remaining parakeet (*Pyrrhura vittata*) was fairly abundant during autumn. A characteristic is the quite extraordinarily accurate correspondence of its colour with the chlorophyll green of plants, as was first borne in on me by the experience noted in my diary on 31st March (see p. 55).

Of parrots the most conspicuous was the Blue-fronted Amazon (*Chrysotis aestiva*) abundant in spring, its cries, loud enough to be heard at a quarter-mile distance, resounding in all directions. The cry *caa-caa* is of lower pitch than that of other species and uttered in wonderful variety, amusingly suggestive of human emotions of fright, indignation, remonstrance, according to circumstances. They were most usually in pairs but occasionally in small flocks of a few pairs in company.

Maximilian's Parrot (*Pionus maximiliani*) was for a time, in the autumn, quite abundant, usually in small parties of three or other odd numbers, but later on it became less numerous.

One day (7th January), hunting in the monte, I observed a Maximilian's parrot sitting in a low bush and thought I would attempt to catch it, so I slowed down my movements to the minimum and gradually glided towards the parrot, which kept its gaze steadily concentrated on me but showed merely interest and no alarm until I successfully grabbed it—when the air of the monte at once resounded with its wild shrieks.

Birds of Prey. The open plains of the Chaco with their richness in animal life are under the constant surveillance of birds of prey, varying in size from small falcons like the Cinnamon Kestrel (*Tinnunculus cinnamominus*) to the large Crowned Harpy Eagle (*Harpyhaliaetus coronatus*). Most frequently it would be the characteristic voice that drew one's attention to a bird of prey perched at some elevated view point; the loud shrill cries of a crowned harpy perched on one of the upper branches of a dead quebracho; the prolonged shrill whistle of the Grey Crane Hawk (*Geranospizias caerulescens*); the gruff *ha ha* of *Herpetotheres cachinnans*, as it peers at one bobbing its head up and down; the prolonged *ĕh* something like the cry of a lamb uttered by *Busarellus nigricollis*, the beautiful eagle mentioned on p. 58 with plumage chestnut in colour apart from the creamy white head, long black wings, and a very graceful flight.

Particularly attractive to the birds of prey are the great fires which, sweeping through the vegetation, turn out innumerable animals from the lairs in which they lurk invisible during the day or kill those unable to escape. At such times one sees soaring in circles the beautiful Chilean Eagle (*Geranoaetus melanoleucus*) with its black breast and white abdomen, or still more frequent the Black Buzzard (*Urubitinga zonura*).

While most of these birds of prey are after the various animals fleeing from the fire, others are interested rather in the carcasses of dead animals left behind: such as the Carancho which abounds in the Chaco as on the Pampa. Still others affect still other diets: such is the case with the Marsh Hawk (*Rostramnus sociabilis*), which as its name implies shows the unaccipitrine habit of going in flocks, and feeds on the common fresh-water crabs and snails. As already mentioned the cream-headed eagle also feeds usually on crabs—although its long sharp talons and the very rough under surface of its feet suggest rather a purely fish diet. On the whole the diurnal birds of prey were catholic in their taste: all seemed to be grist that came to their mill—fish, frogs, lizards, snakes, birds, mammals including bats.

Of the nocturnal birds of prey the most conspicuous was the great Ñacurutu Owl (*Bubo virginianus*) and it contributed one of the most characteristic factors to the night sounds of the Chaco in its deep *boo-hoo-hoo*.

Not so abundant and much less conspicuous was the little Pygmy Owl or Caburé (*Glaucidium ferox*). When it ventures out by day it is mobbed by all the birds of the neighbourhood and our soldiers told us picturesquely how it can be seen in the late afternoon sitting in a tree uttering cries and attracting to itself as if by hypnotic influence all the small birds of the neighbourhood. Eventually the caburé seizes one of the plumpest; the spell is broken, and the assembly disperses, their terrified accents giving place to a paean of praise for their delivery. However, I had already learned by experience that our Spanish American friends tend to be picturesque rather than accurate when they convey information on natural history.

In the monte a third owl voice—*whoohoohoo-hoo-hoo*—was to be heard, emanating from *Scops brasilianus*, intermediate in size as in frequency between the ñacurutu and the caburé.

During the hours of daylight the open palmar is under constant supervision by the two species of vultures so-called (*Cathartes aura* and *C. atratus*), which with their relative the Condor of the Andes constitute a group of birds quite distinct from the true vultures of the Old World. They are to be seen soaring at a great height, each ceaselessly scanning the area of ground underneath, while at the same time keeping a close eye on his fellow-vultures, ready to hurry after anyone of them who by accelerated movements betrays the presence of something of common interest such as a dead or dying animal. If one shoots a deer and neglects to cover it at once with grass the vultures will soon arrive in streams and within half-an-hour the animal is reduced to a clean skeleton. They will not however attack so long as there is any sign of life. Waking from sleep on the ground one occasionally found a vulture or two on neighbouring trees, head on one side and scrutinising one with interest and, no doubt, hope.

Among the small perching birds of the Chaco are representatives of the main groups familiar in Europe. A couple of Thrushes (*Turdus leucomelas* and *T. rufiventris*), the latter of which was observed feeding in an unthrushlike manner hopping after the small fish and prawns in a shallow pool; a Wren (*Troglodytes furvus*); three species of Swallow (*Progne chalybea*, *Hirundo erythrogastra*, *Tachycineta leucorrhoa*); nine species of finches including the two Cardinals (*Paroaria cucullata* and *P. capitata*) with their chaste grey and white plumage and gorgeous scarlet crest.

But more interesting to the European naturalist are those birds which represent groups peculiar to the New World, such as for example the common little yellow-breasted and blue-backed *Parula pitiayumi*, *Geothlypis velata* also with yellow breast but green above, and *Basileuterus auricapillus*, another yellow breast with greenish yellow back—all belonging to the group Mniotiltidae.

Most numerous and most varied, except in their monotonous, predominantly brown colouring, are the insectivorous Dendrocolaptidae—the Wood-hewers and their allies. Perhaps the most conspicuous in the Chaco is the large Chestnut Wood-hewer (*Xiphocolaptes major*), about 12 inches in total length, which abounds especially in the open woodlands, where one may see it advancing vertically up a palm stem in a series of hops; occasionally varying the performance in a startling manner by suddenly dropping a foot or two like a stone, instantly arresting its fall by catching hold of the smooth surface with its claws. Normally it feeds on such insects as it can collect from the bark but on occasion may be seen feeding on the ground, progressing by hops, and digging out insects from the soft soil or fallen leaves with its bill.

It is very tame, letting one approach quite close, and like other Pilcomayo birds very inquisitive; so long as one remained motionless it would hop from twig to twig all round one, watching the while with scrutinising gaze.

Occasionally it may be seen sitting vertically on a tree trunk, propped up by its stiff tail feathers, keeping its mouth wide open and emitting a prolonged creaking sound ending up suddenly with an abrupt ejaculation of lower pitch—*eh-uh*. At other times it emits a series of double notes with gradually falling inflection, while at still others it produces a short but sweet song beginning with low faint notes of considerable variety, a series of these low notes continuing sometimes to the end.

Still more abundant is the little Wood-hewer (*Picolaptes angustirostris*) which at times is to be seen on almost every tree trunk, working round it spirally upwards in a succession of hops, prying into every crevice, probing with its long slender bill, and fishing out its insect prey, mainly small beetles. As it sets to work it emits a short rippling scrap of song, or sometimes a succession of shrill whistling notes in gradually falling cadence. It shows no fear of man, and as it hops round one it presents a curious appearance owing to the axes of its eyes being so slightly divergent that it looks at one full face with both eyes at once.

The name wood-hewer is no longer appropriate to the remarkable *Xiphorhynchus lafresnayanus* (Plate VIII (*a*)), another common and conspicuous Dendrocolaptid, for in it the enormously long and slender bill is useless except as a probe and forceps. When it had gripped a small beetle between the tips of its bill I was able to watch it slide the bill down against the edge of a branch so as to push the insect within reach of the short tongue and so enable the act of swallowing to be accomplished.

It is a Dendrocolaptid (*Phacellodomus striaticollis*) which provides one of the most conspicuous bird's nests in the Chaco, especially in the brush bordering the river—an enormous structure built of dry twigs at the tip of a tree branch which it weighs down, sometimes nearly to the ground. It is not merely the remarkable nest that draws attention to this bird: still more is one attracted by its vocal performance—a duet performed by the male and female as they sit facing one another. One of the two utters a few sharp call notes, the other replies. The first then begins a succession of clear hard notes—at first distinct then becoming shorter and more rapidly repeated until at last they become fused together in a tumultuous rush. The partner joins in a few notes behind the beginner and accompanies it to the end of the duet.

The Icteridae—which take the place of the starlings of the Old World—are again a striking feature in the bird life of the Chaco. They too have a characteristic scheme of colouring, the male birds having black plumage, often diversified however by patches of yellow or red or, more rarely, white.

The long pendant flask-shaped nests of the Icterid *Amblycercus solitarius*—with its uniformly black plumage and white bill—and of *Cassicus albirostris*, also with white bill but its deep black plumage sharply varied by bright yellow rump and wing coverts, have already been mentioned (p. 54). Both species were very abundant and conspicuous—alike to the eye and to the ear.

Then there were three species of *Molothrus*—of which one (*M. bonariensis*) is the common cowbird of the Pampa often to be seen feeding on the backs of cattle and horses and well known from its habit, like that of the European cuckoo, of freeing itself from the responsibilities of incubation and feeding its young by depositing its eggs in the nests of other birds.

Several Icterids frequented especially the beds of reeds and bulrushes in the swamps. Pre-eminent among these was *Amblyrhampus holosericeus*, our old friend of the Paraná delta, conspicuous by its head and neck of brilliant scarlet contrasting with the deep black of the rest of its plumage, and also by its beautiful and plaintive whistle of two notes—the second prolonged and of lower pitch. *Agelaeus ruficapillus* has a similar scheme of colouring but in it the red on the crown of the head and on the front of the throat is a subdued chestnut.

An Icterid (*Icterus pyrrhopterus*) frequently seen in numbers in the tops of the palms is recognisable by the patch of chestnut on the shoulder. As in the case of the first two Icterids mentioned the nest hangs from the tip of a branch but is constructed of lichens instead of long vegetable fibres.

The last of the Icterids to be mentioned is *Aphobus chopi* with uniform black plumage, particularly frequent in July in the open palmar round Fortin Page.

Corvidae. Perhaps the most conspicuous of all the bird inhabitants of the patches of monte was the Urraca Jay (*Cyanocorax caeruleus*) with its boldness and inquisitiveness, and its beautiful plumage of blue and yellow. One would often be surrounded by a mob of these birds hopping from branch to branch, approaching sometimes to within a yard or two, and behaving in general as I have described on p. 52.

That the bird is by nature bold and readily tamed was illustrated by one with a broken wing which I snipped off, for the very next day it showed itself to be quite at home in my hold, hopping about on my knees and shoulders as I sat at work, trying to tear pieces of flesh off the bird on which I was operating, and taking food and drink from my hand without showing the slightest fear.

The other Jay (*Cyanocorax azureus*) with plumage of uniform dark blue, approaching black in the head region, shows curious differences in its manners. In what appears to be the normal condition it is exceedingly shy and it is difficult to approach within gunshot. Very commonly however in the Chaco one finds it associating with *C. caeruleus* and in this case its manners are entirely assimilated to those of its relative, there being nothing then to choose between the two in boldness and inquisitiveness. Azure jays were however usually the first to notice my approach and would at once raise the alarm with their harsh *caa-caa-caa*.

Tyrannidae. A characteristically New World family of birds is that of the Tyrannidae which take the place of and resemble in habits the flycatchers of the Old Word. Of these some fifteen species were to be found round Fortin Page, several of them old friends of the Pampa such as the bien-te-veo (p. 12) and the scissor-tail. Like the larger birds of prey the Tyrannidae are attracted by fires in the palmar, the attraction in this case being the swarms of insects escaping from the flames. Typically the tyrant birds were to be seen perched on the summit of a bush or palm stem from which they made sallies after insects: on the other hand some of them ran about like thrushes, or quartered the ground like falcons.

In point of numbers the brown Short-winged Tyrant (*Machetornis rixosa*) came first in the Pilcomayo region, being abundant along the margin of the water, feeding in pairs, each pair jealously guarding its own territory and violently resenting the intrusion of other birds of their own species.

In colouring, first place goes to the Churinche (*Pyrocephalus rubineus*), its brilliant scarlet under surface, neck and crested head contrasting with its sombre almost black back, wings and tail. At times it was frequent. Other tyrants peculiar in their coloration were the Widow Tyrant (*Taenioptera irupero*)—dazzlingly pure white, apart from the black wings and tail tip—frequent and conspicuous in open spaces perched on the top of a bush waiting for its insect prey, and causing one to wonder whether the bright white colour may not serve as an attraction to flying insects much as does a lighted lamp. Other black and white members of this family of birds were *Fluvicola albiventris*, not common but occasionally to be seen hopping about among the drift wood by the river's margin, and *Arundinicola leucocephala*, black with white head and neck, frequenting reed beds but rather rare. *Euscarthmus margaritaceiventris* with dark head, olive green back, and pearly white belly was often to be seen among the brush by the river.

Another of the characteristic New World families of birds is that of the Cotingidae. Only two of the many known species were met with on the Pilcomayo but these included one of the most beautiful—a little Greenlet (*Pachyrhamphus viridis*), olive green above, pale creamy buff below, bright yellow breast and satiny black cap—the colour harmonising perfectly, with no suggestion of gaudiness.

Of that other great New World group of birds the Tanagers, closely allied to the finches both in structure and habits, only ten species were met with on the Pilcomayo. Of these, three at least were conspicuous by their brilliant colours—*Euphonia chlorotica*, dark purple and brilliant yellow, *Tanagra bonariensis*, blue and yellow, and the beautiful scarlet *Pyranga azarae*. Beautiful too in its more sober colouring was the blue-grey *Tangara sayaca*, abundant during the winter months hopping about among the twigs of trees and bushes in company with other tanagers and finches. *Saltator caerulescens* also was one of the commonest birds among the bushes and scrub.

It will already have been gathered from my diary (p. 58) that a highly characteristic element in the bird fauna is contributed by the Bush birds belonging to the New World family Formicariidae, conspicuous as they hop about in the brushy undergrowth in their black and white speckled plumage; i.e. the male birds, for the plumage of the female is of the earthy chestnut so common among the small birds and no doubt obliterative in its significance. The Bush birds seen on the Pilcomayo all belonged to two species *Thamnophilus major* and *T. radiatus*. Of the two the latter was particularly bold and inquisitive, hopping round and eyeing one curiously and presenting rather a comical appearance with its black crest elevated. Both had characteristic voices, in the case of the greater bush bird a succession of sharp almost chuck-like notes at first quite distinct but gradually becoming more rapid until they passed into a confused trill and ending up with a miauw-like sound—*chook*, *chook*—*miauw*. In *T. radiatus* the song *whoo-whoo-whoo* consisted of a low clear note repeated three or four times ending in a slight trill.

I came across a third species (*Formicivora strigilata*) in the Mushcui country farther north but never on the Pilcomayo. The Mushcui regard it as a bird of evil omen and believe that its cry sets on fire their shelters and blankets.

Caprimulgidae. Of the three species of goatsucker encountered in the Chaco two (*Podager nacunda* and *Hydropsalis furcifera*—the Scissor-tail Goatsucker) were commonly flushed when nearly trodden upon as they lay on the bare ground by the margin of the river or swamp, masked effectively by their extraordinarily perfect obliterative colouring. They were usually in small flocks of five or six individuals and fluttered away for a short distance, then dropped suddenly and became again invisible. The third species, the well-known Whip-poor-Will of North America (*Chordiles virginianus*), has a no less effective obliterative colouring but when at rest was to be found crouching close down on the surface of a branch of a tree in the monte.

The family of cuckoos was represented by a group of birds strangely unlike the cuckoo of Europe. Most conspicuous were the two species of *Crotophaga*, black with a metallic gloss and characterised above all by the strange beak expanded into a vertical plate taller than it is long. *C. ani* the smaller of the two abounded all along the Pilcomayo but especially where there was open brush. Their frequently uttered cry a clear pipe something like that of the curlew. The larger species *C. major* was less common and seen only during the summer. Its cry, a short *ĕh* rather like that of the Eagle *Busarellus*, varied sometimes by a series of rapidly uttered somewhat whistle-like sounds. Next in abundance to the Ani was the brown *Guira piririgua* commonly miscalled by the Argentines 'Urraca' (magpie). Another strange-looking Cuckoo, *Piaya cayana*, with deep chestnut plumage and a disproportionally long tail, was to be found feeding on the ground in the dense forest. At night in the forest one occasionally heard the sweet dissyllabic cry of *Diplopterus naevius*, the 'Wuchen' of the Tobas, the possession of even a fragment of which brings them as they believe good fortune.

In the Chaco with its great expanses of estero or swamp a conspicuous part of the avifauna is constituted by those groups of birds that are specialised as feeders on fish. Of kingfishers there are three species of the New World genus *Ceryle* (*C. torquata*, *C. amazona* and *C. americana*), all abundant and to be seen perched on branches overhanging the river or at times hovering hawk-like over the water and descending in a steep dive in the, usually unsuccessful, attempt to secure a fish. The small *C. americana* as it sits on its perch utters a sharp *tic* at intervals and occasionally sounds its call-note—a subdued but very high-pitched prolonged whistle which brings an answer from another bird. In the case of *C. amazona* one may sometimes see the male sitting on his branch with wings expanded and body nearly vertical indulging in a regular song not unlike that of the oven-bird and consisting of a rapid succession of chirpy notes.

Again one may often see perched on a branch overhanging the river a Cormorant (*Phalacrocorax brasilianus*) or a Darter (*Plotus anhinga*). The cormorant though typically a sea bird was abundant in the Chaco and noteworthy in the fact that it remained there right through the dry season when the waters of river and lagunas had become intensely salt.

Ardeidae. Herons were abundant and their generally well-fed condition belied the old-fashioned idea that their attenuated form of body is a result of poor nutrition. It is rather to be associated with the assumption of an obliterative attitude favoured by various species with neck and beak stretched rigidly vertically upwards, and which has given rise to the popular name Mirasol or Sun-gazer. The efficacy of this protective posture is increased as already indicated (see pp. 9 and 42) by the obliteratively coloured back being turned towards the potential enemy. The five species met with on the Pilcomayo included the two beautiful white Egrets (*Ardea egretta* and *A. candidissima*)—both in abundance except during the period of greatest drought. The Whistling Heron (*Syrigma sibilatrix*) with its hard metallic notes like that of the Bandurria was one of the excellent exponents of the obliterative attitude I have mentioned. A Bittern (*Tigrisoma marmoratum*) was frequent wherever there was fresh water, especially among the beds of tall bulrushes from which when disturbed it would rise up uttering a loud and hoarse *ha-ha-ha*.

Of the stork family three species were conspicuous in the Chaco —the Maguari Stork (*Euxenura maguari*) already familiar in the regions farther south, the great Jabirú (*Mycteria americana*) and the 'Wood Ibis' (*Tantalus localator*). The most striking of these is the jabirú, standing 5 feet in height with immense wing spread, pure white plumage apart from the neck where the skin is naked and black except at its base where the dilatable crop region is encircled by a scarlet band—much less brilliantly coloured in the smaller female. As we steamed northwards towards the Pilcomayo jabirús became a familiar sight, standing about on one leg in flocks, widely dispersed, the very picture of unsociability. On the Pilcomayo they were fairly plentiful, often seen flying overhead or standing about on their favourite haunt—a flat sand-spit by the river or laguna margin.

27th July: A small lakelet of dark still peaty coloured water reflecting as in a black mirror the surrounding palms and other trees: at one end a patch of tall bulrushes ten feet or so in height with their brown velvety flower-spikes. And the most striking part of the picture—a great Jabirú, standing on one leg, its plumage white as snow, its black neck curled up on its shoulder, its heavy beak pointed downwards resting upon its breast, motionless and silent, pensive as if it were the philosophic genius of the place.

The wood ibis, so miscalled from its curved beak resembling

that of an ibis, was the least common of the Chaco storks, usually to be seen in the distance soaring in great circles over a camp fire.

Of the four species of true ibis met with on the Pilcomayo the most conspicuous was the Bandurria (*Harpiprion* or *Molybdophanes caerulescens*) frequently seen in small flocks passing overhead or soaring over camp fires, or probing the bare ground for food by the margin of a laguna or in an area over which a camp fire had passed. When disturbed they will fly up and alight each on the summit of a dead palm, uttering their loud and characteristic cry —a high-pitched abrupt metallic note repeated in quick succession —one of the characteristic voices of the Chaco and which incidentally gives the bird its popular name in Argentina from its resemblance to the twang of the stringed instrument of Spain, the Bandurria.

Another Ibis (*Theristicus caudatus*) shares the name Bandurria, for it too has a voice of hard metallic notes. During the winter it was to be seen soaring in flocks over burnt areas and it occurred in lesser numbers all through the year although it belongs more characteristically to regions farther south.

The White-faced Ibis (*Plegadis guarauna*) was occasionally to be seen feeding by the margins of the swamps but less abundant than the two species already mentioned. The specimens I shot were distinctly smaller and more brilliantly coloured than those I had collected on the Pampa.

Finally the Whispering Ibis (*Phimosus infuscatus*) turned up very occasionally.

The species of spoonbill and flamingo met with on the Pampa occurred also in the Chaco—the former more frequently.

The Chajá (*Chauna chavaria*), that strange rather old-fashioned bird as big as a goose, often to be found wading about in shallow water and sometimes swimming, was throughout my period in the Chaco always to be seen in isolated pairs whereas in the south, e.g. in the region of the Paraná delta, they were congregated in large flocks. If anything happened to arouse their suspicion while feeding the chajás would at once fly off, uttering their loud cries, and alight on the summit of a dead palm or other elevation with an unobstructed view all round. At night they roosted in similar situations and were extremely vigilant, at once raising the alarm if any large animal moved about in their vicinity. We found them useful as sentinels when unfriendly Indians were prowling about at night.

Owing no doubt to the comparative failure of the wet season during our year on the Pilcomayo the family Anatidae which includes swans, geese and ducks was poorly represented as compared with what I had seen on the Pampa. The only duck really common was the Muscovy (*Cairina moschata*) usually occurring singly or in small parties but occasionally in flocks of as many as twenty. Three species of Teal (*Querquedula cyanoptera*, *Q. versicolor*, *Q. brasiliensis*), two of Tree-ducks (*Dendrocygna fulva*, *D. viduata*), a Shoveller (*Spatula platalea*) and a Pintail (*Dafila spinicauda*) completed the Pilcomayo list. In the Mushcui country, however, occurred also the beautiful large black and white Duck *Sarcidiornis carunculata*.

The large Picazuro Pigeon (*Columba picazuro*) was common on the Pilcomayo, being seen most usually winging its way rapidly overhead, or resting on the topmost twigs of trees or on dead palm stems, uttering a soft cooing song which commencing imperceptibly swells up rapidly to its full volume—*he-oo; coo-coo-caoo; coo-coo-caoo; coo-coo-caoo*. At other times they utter a series of abrupt *coo-hoo*'s. Very shy of man.

In much greater numbers were the smaller Doves (*Columbula picui* and *Chamaepelia talpacoti*) and also the Solitary Pigeon (*Engyptila chalcauchenia*).

In the Chaco I became acquainted for the first time with the Cracidae or Curassows—a characteristic New World group of fowls. The three Pilcomayo species, the Mitu (*Crax sclateri*), the Yacu hu (*Penelope obscura*) and the Charata (*Ortalis canicollis*), have been already sufficiently referred to (see pp. 40–51). Of the three only the last mentioned was common in the Fortin Page neighbourhood and as already indicated its voice provided one of the most familiar Chaco sounds.

Of three species of water rail the Paca-á (*Aramides ypecaha*) has also been fully dealt with. Its voice, too, contributes one of the characteristic Chaco sounds (p. 56). Still others were provided by the piercing shrieks of the Carau (*Aramus scolopaceus*) and the wild cries of the Chuña (*Cariama cristata*, p. 87). Both of these birds belong to distinctive New World groups and both were common about Fortin Page.

Of the plover family the Teru-teru was not nearly so abundant in the Chaco as on the Pampa, but the charming little Azara's Plover (*Aegialitis collaris*) was often to be found on the bare margins of salt lagoons. It fed on small molluscs, insect larvae, etc. (see also p. 61).

An occasional Stilt (*Haemantopus brasiliensis*), a Snipe (*Gallinago paraguiae*), the Pectoral Sandpiper (*Tringa maculata*), the Greater and Lesser Yellowshank (*Totanus melanoleucus* and *T. flavipes*), and the Solitary Sandpiper (*Rhyacophilus solitarius*) were other representatives of the great plover family in the Chaco.

Of grebes—probably again owing to the absence of the normal rainy season—the five species seen on the Pampa were on the Pilcomayo reduced to two (*Podiceps rollandi* and *Podilymbus podiceps*) and even these were rarely seen.

As regards the characteristic New World game-birds belonging to the Crypturi or Tinamus the Martineta was common and its plaintive call was a familiar sound in the palmar during the springtime, while another species (*Crypturus undulatus*) was found only in the deep monte.

Of reptiles there come first the two species of alligator or jacaré—the larger *Caiman sclerops* growing commonly to a length of 7 or 8 feet and the rather smaller *C. latirostris*. During the wet season these were widely distributed through the swamps and lagoons but as these dried up the jacarés concentrated along the larger rivers. In long stretches of the River Paraguay during the dry season there would be a jacaré every 5 yards or so, and here and there groups of four or five close together. Many however fail to escape from the drying-up swamps and these take refuge by burying themselves underground after the fashion of so many other swamp-frequenting animals. The word alligator is, as is well known, a corruption of the Spanish 'el lagarto'—the lizard—and it is interesting to find sometimes in a modern Spanish book the old habit followed and the alligator actually referred to as 'el lagarto'. The flesh of the jacarés' tail is used by the Indians as food and is as already indicated quite agreeable.

Of tortoises two were quite common, the larger Brazilian Tortoise (*Testudo tabulata*) and a smaller one called Pōtrunrut by the Indians (? *T. argentina*).

Of poisonous snakes there was a considerable variety the two most formidable being the Rattlesnake (*Crotalus terrificus*), very common in the Chaco, and the Vivora de la Cruz (*Bothrops alternatus*). These poisonous snakes do not normally attack human beings but if cornered they settle down into a coil ready to strike. The rattlesnake produces its whirring noise through portions of its horny epidermis which remain loosely attached to the tip of the tail when the rest is shed. It was of special interest to note that as already mentioned the Vivora de la Cruz, although not provided with such a rattle, foreshadows the evolution of one by vibrating the tip of its tail rapidly amongst the vegetation, producing in this way a distinctly audible warning sound.

In my experience these poisonous snakes were roused to action only by sudden movement. Gliding gently towards the snake without any sudden movement it was always possible by shooting out one's hand suddenly, to grip it behind the head before it had time to arrange itself in the coil necessary to strike. I noticed that non-poisonous snakes on the other hand in similar circumstances shot out their heads, warning one off by a feint attack while the poisonous snake confined itself to the real thing.

By far the most interesting thing however connected with the snake fauna of the Chaco is the discovery of the former existence in that region of gigantic poisonous serpents of dimensions previously undreamt of. One day, long after my return home, I had a visit from my missionary friend Andrew Pride, on leave from his work among the Mushcui Indians. He announced that he had a present for me and putting his hand into his pocket produced what he called a 'claw' of some animal which he had found amongst a mass of bones exposed in digging a trench near his Mission Station. I had only to look at the specimen to see that it was no claw but something far more interesting, namely the poison fang of a gigantic serpent (Plate VIII (*b*)). Down one side ran a deep groove for the transmission of the poison, precisely as in some of the more primitive poisonous snakes of the present day (*Opisthoglypha*) in which the poison groove has not yet become closed in, as it has in the more highly evolved snakes, to form a tube running the length of the fang and opening close to its tip.

In accordance with accepted rules I had to coin a new name for this strange new type of extinct monster so I called it *Bothrodon* from βόθρος a trench or furrow and ὀδών a tooth, dedicating it to Pride, its discoverer, as *Bothrodon pridii*.

In view of the extraordinary interest of the specimen I impressed upon Pride that the remainder of the skeleton simply *must* be found. Alas! careful and prolonged search failed to discover the exact point where the bones had been reburied in the soil and so *Bothrodon* remains and may for ever remain represented by the solitary poison fang which now constitutes one of the treasures of the Museum of Zoology of the University of Glasgow.

In the absence of the complete skeleton it is of course impossible to give an exact estimate of the total length of *Bothrodon*, but comparing the size of the fang with that of a modern poisonous snake such as the rattlesnake (Plate VIII (*b*)) it may well have been some 60 feet, more than four times the length of the longest modern poisonous snake.

Comparison with the fang of a typical modern snake brings out another point of special interest, namely, the more marked curvature, the more hook-like form, of the *Bothrodon* fang, suggesting that its function was not merely to strike but also to hang on to its prey. In imagination one may picture the grim scene of one of the great contemporary mammals such as a Megatherium or a Mylodon dragging along the vast bulk of the hideous monster until overcome by the poison seeping into its body.

The only serpents of great size existing at the present day in the Chaco, or indeed anywhere else, are the non-poisonous boas and their allies which kill not by biting but by encircling their prey in coils of their body and crushing the life out of it. The largest of these is the Anaconda (*Eunectes murinus*) which is said to reach over 30 feet in length. The biggest specimen in my Chaco experience appeared one day when I was paddling alone in a canoe. It was swimming but in spite of repeated blows on the head from my paddle it managed to escape. It might measure nearly 20 feet in length so far as I could judge. As in the case of other reptiles, such as jacarés and lizards, we found the flesh of these large serpents to be quite good eating.

As regards the great variety of lizards encountered on the Pilcomayo I need only mention *Tupinambis teguexin*, the large 'Iguana' of the Argentines already met with in the Pampa; and the small species included in the lichen-like fauna of the tree tops. Occasionally too we dug up an Amphisbaenid—one of these remarkable lizards which are adapted to burrowing in the ground and have taken on the appearance of earthworms, the body being long and slender and the limbs having completely disappeared.

As already indicated the heavy and long-continued rains of a normal wet season did not occur in the summer of 1890–91, but even the occasional showers served to disclose the existence in the Chaco of a great variety of frogs and toads. At such times during the night the air was filled with a really marvellous concert of frog voices. My expedition to the Chaco six years later was favoured with a rainy season even greater than the normal and my companion Budgett was able to make an intensive study of the frog fauna so I will reserve what has to be said about it for Chapter IX. I will also defer what has to be said about the fishes of the Chaco.

One of the chief torments of the Chaco comes from the superabundance of insect pests. First among these are the mosquitoes of which there are many species differing greatly in size and habits. During the warm season they exist in enormous numbers, different sets succeeding one another throughout the twenty-four hours though of course most numerous during the hours of darkness. Riding along through the Chaco the head of one's mule would often appear as if covered with staring fur—composed of mosquitoes in close array, so that one had to keep a switch of twigs in constant operation to disturb their blood-sucking. When camped for the night we would make a separate smoky fire for the mules and they would crowd into its protective smoke. As already mentioned, when collecting in the early morning one would be surrounded by a cloud of mosquitoes, and often escorted by a few dragon-flies busily hawking at the mosquitoes during flight.

Among the mosquitoes were many Anophelines recognisable by their attitude when biting, and no doubt responsible for our frequent malarial attacks though their role in this connexion was in these days as yet unknown.

An alleviating feature of the mosquito nuisance was that, bitten as we were practically continuously when outside the shelter of our mosquiteros, we soon developed complete immunity towards the irritant poison secreted by the insect and injected into its bite so as to increase the blood flow to the particular point.

Such alleviation did not apply in the case of another important pest—the Midges (*Ceratopogon* or *Culicoides*) which every now and then appeared in myriads. While always much smaller than mosquitoes, some of the Chaco midges were so minute as to be barely visible to the naked eye and on that account were called polvorinos by our Argentines. Against these the ordinary, relatively wide-meshed, mosquitero is no protection: it has to be made of cambric or similar fabric to keep them out. When the plague of polvorinos was exceptionally bad we took refuge in the river, sitting completely submerged but for pushing up our noses to breathe.

The advantage of these midges not being so continuous in their attentions as the mosquitoes was far more than offset by their failure to induce immunity to their poison. In view of the irritation produced by the infinitesimally small amount of the midge poison injected when it bites, that poison must be of quite extraordinary virulence.

There were many other insect pests less formidable than the mosquitoes and midges; chief amongst them large Horseflies (*Tabanus*) and the Viuda or Widow Fly (*Chrysops*) so called from the wings, dark except the transparent tips, suggesting the appearance of a mourning cloak. The Viuda was particularly light-footed, an adept in alighting undetected on one's skin and in slipping out of the way of an attempted slap.

An interesting experience came when I was consulted about a large lump which had developed on the back of one of our soldiers and which when opened displayed a big larva of the Warble Fly (*Hypoderma*) well known for the injury it does to hides. Quite a number of subsequent cases turned up among the men who were often working with the upper part of the body exposed.

The occasional plagues of the ordinary flea are mentioned elsewhere (see p. 178) but there also occurred in the Chaco the troublesome Jigger (*Pulex penetrans*). In this case the female after fertilisation gives up her active habits and burrows through the skin of the victim, usually of the toe. Here she remains in peaceful seclusion, absorbing nourishment and gradually growing to about the size of a pea, while she continues the process of extruding her eggs. Ordinarily no discomfort is felt, only a slight itching. Removal of the jigger by working a needle round it may however cause trouble if the wound is allowed to become septic. In the case of our soldiers, keeping on their top boots for a prolonged period, the trouble became really serious through the eggs passed out by the female jigger developing within the boot and thus proceeding to bring about multiple infection of the foot. The dog was the only other animal in which I found jiggers and no doubt they play an important part in keeping up the infection of man.

As will have been gathered from my diary the abundance of insect life in general is one of the great features of the Chaco. In certain parts the ground is thickly studded with ant-hills, each inhabited by a large community of termites or white ants. The normally invisible members of these communities constitute the staple food of various insectivorous creatures such as the great ant-eater or the ground-frequenting Woodpecker (*Colaptes agricola*). But when swarming takes place and the winged mature termites emerge in their millions preparatory to starting off in flight to found new communities this is the signal for an immense concentration of insectivorous birds of all kinds which gorge themselves to repletion upon the termites swarming over the ground everywhere.

Usually the ant-hill houses one or more communities of true ants in addition to the termites. Thus in one ant-hill which I opened there were present first the large termite community, secondly a community of large black ants, and thirdly a community of small red ants—each community inhabiting a separate portion of the ant-hill.

Within the monte, a common species of termite was one which constructed a large spherical nest of masticated wood attached to the bole of a tree some several feet from the ground. Termites are creatures of the dark; the cause of the white appearance which gives them their popular name of white ants is that their skin is devoid of the dark pigment so usual in many different types of animal to protect the delicate living protoplasm from the harmful effect of daylight. For indeed daylight is inimical to the activities of living protoplasm, although in the case of man this harmful influence is apt to escape notice in face of the beneficial effects on health produced in dark and gloomy temperate climates by the psychological influence of bright sunshine when it does occur. The particular species of termites just mentioned which have forsaken their normal subterranean existence and taken to forming their nest on a tree some distance from the ground have accordingly had to safeguard themselves by forming a light-proof mode of access through a small tunnel built of the same material as the nest and leading to it up the surface of the tree-trunk.

The occurrence of well-marked seasons in the Chaco has had an interesting result in that certain wasps have developed the habit, usually associated with bees, of storing food in the form of honey to tide them over the flowerless winter season. On the Pilcomayo two such species were conspicuous, *Nectarina lecheguana* with large spherical nests and the smaller 'Bala'—probably a species of *Polybia*. In both of these the nest is of the paper-like material formed of masticated wood familiar in the nests of some of our common European wasps.

Of honey-collecting bees there were also two abundant species, *Melipona favora* var. *Baeri*, called by the Argentines Moro-moro, which lives in hollow trees and stores its honey in large wax receptacles, and ?*Emphor fructifer*—Alpamisqui of the Argentines, whose nest is underground. Both of these belong to the group of small 'stingless' bees but they are not without means of defence. When riding through the forest if the leading mule passed close to a nest of Alpamisqui the owners would at once sally out and proceed to attack and bite mules and their riders viciously. When this happened the mules would scatter in all directions and it was necessary to flatten oneself down on the animal's back to avoid being brained as the mule dashed under overhanging branches.

The true ants, present everywhere in enormous numbers and great variety of species, caused us little trouble. On one occasion while still inexperienced and reconnoitring on the bank of the Bermejo I climbed a little mound to obtain a better view and was soon appraised by sharp bites all over my body that the mound in question was an ant-hill. And again, as will be related in Chapter IX, there is the liability, when asleep on the ground, of being attacked by an army of foraging ants. When on watch one night on the Pilcomayo I was shocked to find our mooring cable being used as a gangway carrying two actively moving streams of ants, those coming on board empty-jawed, while each ant moving in the opposite direction bore in its jaws a grain of rice from our small remaining store.

However the ants on the whole were not an active pest, while they did provide endless variety and interest. They included the leaf-cutting gardener ants (*Atta*) already met with on the Pampa and one of the species encountered at Fortin Page was found to be peculiar inasmuch as the community inhabited a chamber excavated in a tree trunk instead of the usual ant-hill.

Particularly conspicuous on the open grasslands were the grasshoppers and locusts in great variety—some of them showing beautiful camouflage such as the species already mentioned as living among the lichens of the tree-tops. At quite an early stage of the expedition the *Bolivia* became infested with the large American cockroach and as provisions on board became scanty these became rather a nuisance at night by crawling over one and nibbling one's hair and nails.

Among the great variety of spiders in the Chaco one stood out pre-eminent in interest, *Aranea socialis*. Walking or riding through the palmar in the early summer one frequently encountered strong horizontal threads of yellow silk two or more in number, parallel, one over the other and stretching between two palms 30 or more feet apart. These were the foundation lines for a communal arrangement of webs belonging to the above-mentioned spider. The owners, a community of small spiders, black with conspicuous scarlet markings, were to be found near one end of the arrangement, resting among a complicated network of threads constituting their home during the hours of daylight. As evening fell the community woke up and became a scene of great activity. The first job was to strengthen the silken cables stretching from palm to palm. One behind the other the spiders ran rapidly along the cable on its under side, back downwards, each adding an additional strand as it did so. When the main cables were sufficiently strengthened the spiders spaced themselves out along the uppermost one and did a kind of trapeze act, dropping down to the lower cable and so dividing up the space between into a series of rectangular meshes. Each spider—350 I counted in one case—appropriated one of these. Fixing a thread to one corner it carried it round the bounding thread and fixed its far end to the opposite corner. Repeating the process over and over again it divided the space by a number of radiating threads fixed together in the centre. Tangentially placed threads braced these radii firmly in position so as to constitute a rigid framework, and now followed the final stage, the equipment of the web with its fly-catching device. This consisted of a thread carried from one ray to the next, commencing at the periphery and continued spirally until the centre was reached. This thread differed from its predecessors in that it was coated with a glutinous liquid which rapidly ran into tiny sticky droplets for the entanglement of the insect prey.

Its work completed, each spider settled down in the centre of its web for the nightly vigil. Its legs stretched out radially, gripping firmly the tense radiating threads of the snare, and any quivering of one of these betraying the presence of an intruder, was at once responded to by the owner jerking the thread so as to throw the neighbouring part of the net into violent vibration and thereby increase the chance of the stranger being effectively entangled. The spiders' vigil lasted throughout the night. When daylight came they destroyed the greater part of the web, though leaving the main cables intact, and retired to their daylight homes.

This communal life of the spiders is merely temporary. As sexual maturity comes on a complete change takes place both in habits and in appearance. The conspicuous black and scarlet is replaced by obliterative colouring. The social habit disappears, though there still tend to be a number of webs in fairly close proximity. The large females maintain their industrious habits much as before, sitting in the centre of the web, a beautiful piece of engineering 2–3 feet in diameter constructed of yellow silk and tightly braced up by strong guys.

The male, differing from the female in his smaller size and more slender form as well as in his idle habits, lives apart, skulking diffidently in the neighbourhood of the female's web. Round the margins of the web are usually to be seen small thief spiders on the look out for any crumbs that might fall from the table of their industrious and prosperous neighbour.

Anything of an uneatable nature becoming entangled in the web is rapidly disentangled and cast off, while refuse matters are rolled up in web and arranged in a radial line running from the centre to the periphery.

The warfare of the spiders against the insect world is by no means unreciprocated in the Chaco. Never have I seen such rapidity of animal movement as in an attack upon one of the large so-called Bird-eating Spiders (*Mygale*) by a large black Wasp (?*Pepsis*) in which the latter proved victorious. The largest scale warfare on spiders is, however, carried out by Mason Wasps (*Pelopaeus*), about 1¼ inches in length and of a black and yellow colour, which provide a store of spiders as a supply of fresh food for their grubs. They frequently made use of the shelter afforded by the deckhouse of the *Bolivia* for the location of their tubular nests of hardened mud, attached along one side to the surface of walls, roofs, tables, clothes or almost anything, and it was fascinating to watch their operations. The female wasp would first appear on a reconnoitring expedition. She would fly up and down seeking for a suitable spot, a particular favourite being the rough outer surface of my pith helmet which normally hung on a peg close to the door. Having made her selection she would disappear and presently reappear, this time carrying a large ball of mud in her mandibles. Having attached this she would fly away and repeat the operation time after time, each ball of mud being added to those already in position to build up a tube about two-thirds of an inch in length, closed at the base but open at the top. As each ball of mud was added to the margin of the tube it was spread out by the wasp's tongue and carefully smoothed into shape by the two forelegs, one inside, one outside, the antennae vibrating over the fresh surface all the while as if to detect any irregularities.

Her building operations finished the wasp flew away, returning from time to time to inspect her work. Eventually when she found it completely dry and hard she would again fly away but this time on her return journey she would have a small spider in her jaws—rendered paralysed and helpless by the poison of her bite. This she would stuff into the tube, and, the foraging expedition being repeated over and over again, the nest would be filled to the brim with spiders. In one case I counted as many as ninety. The wasp would now deposit an egg within the nest, seal up the opening, and make her final disappearance, leaving the limp and helpless spiders as a supply of ever-fresh food for the young wasp larva.

There was little time and few facilities for studying the lower forms of invertebrate life in the Chaco but it should be mentioned that an interesting type of earthworm turned up for which a new genus was founded by the then great authority on this group, F. E. Beddard, and named after its discoverer. The discovery of this Chaco species, *Kerria halophila*, was followed by the discovery of about a score of others in widely separated parts of the world and led in due course to the foundation of other allied genera and their designation by new names such as *Kerriona*.

The whole group turned out to be of special interest to students of the anatomy of earthworms and also to students of the geographical distribution of animals as bearing on the possible continuity during earlier geological ages of continents and islands now separated by sea.

The patches of hardwood dicotyledonous forest or monte duro which diversify the open palmar vary in extent from mere clumps of trees to belts of forest several miles in extent, orientated usually in a north-west and south-east direction, i.e. roughly parallel to the general course of the Chaco rivers and streams.

These islands of monte are composed for the most part of small and slender trees not more than 15–20 feet in height, the most abundant species being the Myrtaceous Arrayan or Ñanga Piru (*Eugenia uniflora*) with white blossom and a small red tomato-like fruit of delicious flavour. Along with it are such trees as the Garabato—an Acacia (*Acacia tucumanensis*?) with pale yellow flower-balls and characterised especially by its innumerable recurved spines which make it difficult to extricate oneself when once entangled among its branches, the Tinticaco (*Prosopis adesmioides*), and the Chañar (*Gourliea decorticans*). At intervals the uniform line of these small trees, close together and entwined with slender lianas, is broken by a large forest tree towering above it some 50 feet or so. Such are the Guayacán (*Caesalpinia melanocarpa*) with its smooth green bark and delicate feathery foliage; the Quebracho colorado (*Quebrachia morongii*)—new to science this local species of Quebracho—with its tall and perfectly straight bole and its bark deeply divided into quadroidal portions; and the Wilyik of the Tobas, a tall slender Bignoniaceous tree (? *Tecoma* sp.) with large yellow flowers, whose wood provides the Indian with his fire-sticks.

The interior of these montes is, as a rule, very dry and the closeness of the small trees, the intertwining lianas and the superabundance of spiny plants makes them almost impenetrable (Plate IX (*b*)). Among these spiny plants the most abundant are the two Bromeliaceous plants known by the Guaraní name Caraguatá. One of these is the Caraguatá-i or Water Caraguatá, already mentioned from the fact that its leaves have large sheathing axils in which the traveller may always find an accumulation of cool fresh water—with, by the way, its own peculiar fauna of small crustacea and other microscopic animals. The other Caraguatá—'chaguar' of the Argentines; 'ivira' of the Paraguayans—is the plant which, as mentioned in the next chapter, provides the Indians with the excellent fibre used for spinning into thread and string, of which are formed bags, nets and the fabric used for the men's garments.

These patches of forest are during the dry season highly inflammable. When once set alight they burn furiously, leaving nothing behind but a thick layer of ashes among which scattered burrow-like holes indicate where the highly combustible roots of the forest trees had burned away for some distance below the surface of the soil.

The monte had its own fauna though many of its animal inhabitants are common to it and the palmar. Amongst those peculiar to the monte are of course the monkeys, of which two species are common—the black Howler or Carayá (*Mycetes caraya*) and the Mirikina (*Nyctipithecus trivirgatus*). The latter, first met with on 15th March near the mouth of the Pilcomayo, and provided as already mentioned with enormous eyes fitting it, like those of an owl, for vision in the faintest light, is certainly not in my experience purely nocturnal in its habits as stated by Bates in his *Naturalist on the Amazon*, for when exploring the monte by day I frequently found myself watched by a group of Mirikinas in the topmost branches. Another monte animal was the Cuati (*Nasua rufa*) occasionally seen clambering about in the trees at the margin of the wood.

Apart from the typical monte duro must be mentioned the open woodlands which are commonly to be found along the river margin, where the ground is rendered slightly undulating by deserted loops of the river bed, either completely dried up or still occupied by swamp. In this littoral zone are numerous large trees, almost entirely of two species—the Mandu Virá, unidentified in the absence of flowers, and the Timbo Atá or Timbo Blanco (*Pithecolobium multiflorum*), intermingled with smaller trees such as the Espinillo (*Acacia farnesiana*) and scrub, tangled and spiny bushes, and creepers. It was these woodlands that formed the favourite haunt of some of the wild mammals such as the tapir and the jaguar, and of birds such as the Paca-á.

Finally, there comes a type of scenery peculiarly characteristic of the Chaco, that of the great swamps or esteros. These are a result of the Chaco plain being almost dead level, its extremely gentle slope to the south-east being frequently no more than one inch to the mile, sometimes even less. A secondary result is that the course of the Chaco rivers is extremely tortuous and further that in the regions where slope is least the river channel loses all definiteness, its waters spreading far and wide over the surrounding country. During the long dry seasons this great expansion of the water surface entails an immense amount of evaporation which brings about two results—in the first place a great diminution in the volume of the river between its entry into the Chaco plain from the mountains on the west, and its exit into the River Paraguay on the east, and finally the reduction in area of the water until it is reduced to a tiny streamlet or disappears altogether.

These vagaries of the Chaco waters do much to dominate the animal life of the region. The Indians are affected through changes in the distribution of fish life and game. The Cristianos on the other hand are worried by the blurring of the frontier line formed by the River Pilcomayo which has recently—long after my sojourns in the Chaco—entailed the appointment of an Argentine-Paraguayan Commission to lay down an artificial frontier line through the region in which the river channel is least well defined.

While as regards the lower forms of animal life, those which 'breathe water', or in other words are dependent for their supply of life-giving oxygen on obtaining it from the water in which it is dissolved, are subjected to risk of extermination. In actual fact however, although indeed the drying up of the waters does involve great destruction of animal life yet a large proportion of the swamp fauna manages to survive owing to their having evolved special arrangements which enable them on the one hand to obtain their oxygen directly from the atmosphere and on the other to take shelter for the dry season by burrowing into the mud at the bottom of the swamp, subsisting during their period of imprisonment upon nourishment stored up in their bodies during their period of activity. The Jacaré, the Lungfish, the Swamp Eel (*Symbranchus*) and the large Water-Snail (*Ampullaria*) are typical examples of creatures enabled by such devices to survive the drying up of the swamps. The use of the special air-breathing arrangements is not confined to the period of complete desiccation. Even before this the progress of the dry season is apt to lead to the shrinkage of the water of swamp or river to isolated pools in which the fish population becomes concentrated to such an extent that the water becomes foul and unbreathable, and the fish would be suffocated except in so far as they were able to obtain their oxygen directly from the air. Even under normal conditions in the swamp owing to the absence of currents the deeper parts of the water are extremely poor in oxygen and no doubt this has played its part in favouring the evolution of special air-breathing arrangements.

Altogether the swamps of the Chaco are of fascinating interest to the naturalist. The details of their natural history will be reserved for Chapter IX.

CHAPTER VII

THE NATOKOI

The aborigines of the Chaco are Redskins: they belong to that branch of the human family that until the advent of the white man populated the whole extent of the New World. In different regions they show great differences in detail, e.g. in stature some being tall while others are mere dwarfs such as the Guayakí (Plate VII (*b*)) inhabiting the forests of Paraguay. But such differences are overshadowed by common characteristics which mark them off as one of the main branches or more correctly racial blends of mankind, most nearly akin to the Mongolians, their near neighbours at the northern extremity of the double continent.

Our Indian friends of the lower Pilcomayo, called by their neighbours Natokoi, formed an outlying branch of the great stem known to the Argentines as Tobas and to the Paraguayans as Guaycurús, whose territory stretches far to the west and south-west until it marches with that of the Matacos. The state of guerrilla warfare existing between them and their neighbours had, no doubt, done much to preserve them from the contamination as regards customs and language suffered by tribes having direct intercourse with the white man. While the Natokoi of the Pilcomayo regarded all their neighbours as enemies, it was the Tobas to the south-west and the Nimká to the north-west against whom their hatred was most intense. In a northerly and easterly direction their influence extended much less far, the limiting factor there being tribes of Indians of different stock grouped by the Argentines under the generic name Orejudos from the flat disks of wood worn in the lobe of the ear—to the north and east the Mushcui, of which more will be said in Chapter IX, to the north-west the comparatively small nation of Pihlagá, and beyond them the much more powerful Nimká. The Pihlagá were for the time being in friendly alliance with the Natokoi against their inveterate enemies the Nimká. Both Nimká and Mushcui offered attractive loot to the Natokoi, for they possessed not only sheep, cattle and mules but also plantations of maize, pumpkins and melons.

The men of the Natokoi are tall—several of our friends were over 6 feet—slenderly built and in accordance with their life as hunters and warriors lean and hard: their features fine and regular, and often showing little difference from the Western European type (Plate X). In fact after my return home I was quite startled by the facial resemblance of a former Dean of Westminster to one of the Indian caciques.

Their eyes are deep brown. The difference in stature between the two sexes is very marked—the women being short though slender and graceful in figure.

The dress of the Natokoi consists of a broad rug fastened round the waist-line like a kilt, the two ends overlapping in front of one thigh so as to facilitate freedom of movement. The material is Kaliete or chaguar, the fibre of *Bromelia pinguin* or, less usually, of sheep's wool obtained by raiding the territory of the Orejudos. This kilt-like garment may reach barely to the knee but occasionally forms a fine blanket reaching when doubled right down to the ankles. Such a blanket, half of which can be turned upwards so as to enclose the whole upper part of the body, is highly prized by its owner, and is laid aside during the rough wear of the chase or when fighting (Plate XI (*a*)).

As I got to know my Natokoi better it was interesting to note that some were more 'dressy' than others and that these were by no means the more influential members of the tribe. On the contrary some of those who had most influence such as my friend Chimaki were the least careful about their attire.

Over the main garment of chaguar or wool there is occasionally worn a broad belt of cierbo hide, its lower portion cut into vertical strips which hang freely downwards, so as to allow perfect freedom to the legs when the main garment is laid aside as in a fight (Plate XI (*b*)). During the latter part of my stay with the Natokoi, in which my most devoted follower was Chimaki, I created a special decoration for him in the form of a broad belt of shining tinplate of which he was immensely proud.

The garment of the women is similar in appearance to that of the men but it is shorter, usually not reaching below the knee, and instead of being woven is composed of the skin of the cierbo—the large swamp deer. The skin is dried in the ordinary way pegged out on the ground, and when completely dry is softened by working to and fro between the hands. It is worn with the hairy side inwards and the outer surface is frequently ornamented with not inartistic rectilinear patterns in red pigment.

The toilet of the male is completed by various accessories. Round his waist or more rarely over one shoulder, he carries a small string bag of kaliete to contain his fire-sticks and any little odds and ends that he possesses. His hair—long luxuriant and straight—is cut off at about the level of the shoulder or a little higher and has a broad fringe over the forehead. All of the scanty hair on the body and face is pulled out, and the absence of eye-brows and eye-lashes is somewhat repellent to the European until one becomes accustomed to it.

The Indian takes great care of his hair, combing it frequently with a small comb with long wooden teeth gripped firmly between two pieces of split cane lashed together (Plate XII (*a*)). For cutting it he uses the teeth of the palometa with their sharp knife-like edge (Plate XII (*b*)).

The favourite adornment of the head consists of white ostrich plumes. Part of the hair at the back is tied together to form a small pigtail and to this are attached vertically one, two or three plumes. A similar little tail may be formed over the forehead and occasionally one sees a couple of white plumes passing round behind the cheeks and tied together under the chin. Occasionally again one sees other types of head-dress such as a snood crowned with scarlet woodpecker crests or other coloured feathers, or ornamented with button-like disks of snail (*Bulimus*) shell; but the characteristic head-dress is that of white ostrich plumes. Casques made of jaguar or other skin which occur occasionally have been taken from other nations of Indians.

The white ostrich plumes are used elsewhere than on the head—in the form of armlets and anklets—the latter having also a function other than ornamental, giving warning to snakes or other animals lurking in the grass as the Indian pursues his stealthy way.

Apart from the head a favourite ornament is a necklace composed of rectangular pieces of the shell of the fresh-water mussel (*Anodonta*), worn with the mother-of-pearl surface outwards and forming an effective contrast with the dark coppery red skin.

On ceremonial occasions the face is painted in broad horizontal bands of the bright red pigment 'urucu' derived from the seeds of *Bixa orellana* and obtained by barter from neighbouring tribes.

On the urucu background are drawn narrow black lines made with charcoal along the ridge of the nose and radiating from the angles of the mouth. When on one of the early days of my companionship with the Natokoi I was summoned by Cacique Yordaik to have my face painted up in this fashion I felt that my inauguration into their communal life was complete.

Additions are made on certain occasions to the toilet so far described. One or two of my friends possessed a kind of vest closely woven of kaliete cord and forming an efficient defence against arrows, while the Chief Yordaik possessed a similar garment of jaguar skin. Again when undertaking a long march through rough country, the Indian may bind his feet in moccasins of deerskin, and occasionally one sees rough leggings made of the skin of the great ant-eater.

The women wear their hair as do the men but do not ornament it with feathers. Again they wear necklaces but these are composed of the round disks about the size of a shirt button cut from the shell of the large land-snail *Bulimus*. Each disk has a single hole through its centre and is looped on thin twine in such a way as to overlap its neighbour. Several strings of these are worn hanging loosely round the neck. More rarely one sees necklaces of small round berries strung together.

The women do not paint their faces; they are tattooed. The operation is performed by an old woman who punctures the skin with a plant spine and then rubs in charcoal. The pattern is uniform within the tribe; two parallel lines down the ridge of the nose, a cross between the eyebrows and a pattern of diagonally intersecting lines covering the cheeks and chin. The operation is done normally in two instalments: the two parallel lines and the cross at puberty: the diagonal lines after the first sexual intercourse.

In one or two cases I observed a man with a square of small dots tattooed between the eyebrows but did not succeed in finding out the significance of this.

The equipment of the Natokoi, living in a country without stone, may be said to belong in the main to the Wood Age. His chief weapons are his bow and arrows.

The bow, which in ordinary life is kept strung, is formed of an exceedingly hard, heavy and tough wood which the Argentines call 'cascarandá' and the Tobas 'targík'. It measures in length between 4 feet 10 inches and about 6 feet. Its cross-section in the middle of its length is almost rectangular and measures about 1 inch by $\frac{5}{8}$ inch. Towards each end it tapers gradually, the taper becoming more obtuse near the tip so as to hold the string in position. The outer, convex, side of the bow is noteworthy for two things—its lighter colour, indicating that the bow is cut from the extreme outer layer of heartwood, and the presence of a backstring knotted round the bow at intervals as a guard against over-bending. The presence of a backstring proved of a special ethnographical interest as being the first known instance of such a thing in the New World.

The bow string is usually a twisted thong of cierbo hide although mule hide is greatly treasured when obtainable.

The arrows vary in length from about 3 to about 5 feet, the longer ones being used for fish and for small game, the shorter for fighting and in general for long-range shooting. The shaft of the arrow is a length of jointed cane. The vane, composed of two portions of quill feather bound to the shaft by fibres of kaliete, is normally in one plane but occasionally is arranged spirally so as to cause the arrow to rotate about its long axis during flight. In other words we have here in the case of primitive man a foreshadowing of the principle used in the modern rifle.

The point of the arrow is a piece of targík or kotapík (Quebracho) 15 to 18 inches in length, triangular, biconvex or circular in section; in the case of the last mentioned frequently incised to form barbs and then used particularly for fishing. Other types of arrow-head are used for special purposes. One of these, used especially in hunting the large water-rat or coypu, has a detachable head made of hard bone and possessing a large barb on one side to prevent its withdrawal. The head fits loosely on the shaft and is connected with it by a length of fine twine which unwinds during the flight of the wounded animal and betrays its presence by floating on the surface as it dives under water.

For use against large game such as deer, jaguar, ostrich, the Indian possesses one or at most two arrows in which the head consists of a flat metal blade lanceolate in shape and measuring about 6 inches in length by about $1\frac{1}{4}$ inches in greatest width. These arrow-heads, made from a bit of hoop-iron or beaten-out fencing wire, are obtained by barter through tribes bordering on the white man's territory and are greatly valued. The iron blade is lashed at its base into the split end of a piece of tough wood which in turn fits into one of the ordinary arrow shafts. The blade is kept sharp along its entire edge right down to the base, with the two results (1) that it causes extensive haemorrhage and (2) that it is readily pulled out of the wound as the animal flees away through the vegetation. The Indian is thus enabled to follow the animal's trail of blood and to recover his arrow for repeated use until the chase is brought to an end by loss of blood or by a wound in a vital spot.

In using his bow and arrow the Indian bends his body forwards, his left foot advanced, his left arm stretched out perfectly straight and rigid and the bow inclining upwards at an angle of about 45° towards the right side. The arrow is grasped with the finger and thumb alone.

In attack the Indian can often avoid an arrow coming towards him and it was interesting to note that he would try to escape a rifle bullet by flopping to the ground in reaction to the flash.

The only other usual weapon among the Natokoi is the club (pōn, sulyuranik) made of the densest wood obtainable—cascarandá or palo santo—about 30 inches in length and shaped like a pestle for mashing potatoes. It is wielded in both hands although the weight (about 3 lb.) is sufficient to crash through a human skull with a comparatively gentle touch (Plate XIII (*b*)).

Occasionally but rarely one encounters a lance made of targík. One of elegant design belonging to a rival Chief of the Bermejo Tobas killed by Yordaik and given to me as a keepsake, is depicted in Plate XIII (*a*).

As essential a part of the Indian's personal equipment as his bow and arrows is his fire-drill, composed of the wood of a species of *Tecoma* (Wilyik) (Plate XIII (*c*)). Adjacent surfaces of this wood, owing to the absence of hard annual rings, when rubbed together do not become smooth and polished but, on the contrary, rub down into powder. The drill itself consists of a cylindrical piece of wood 2 or 3 inches in length and about the diameter of a lead pencil. This is inserted into an arrow shaft and can be given a rapid rotary movement between the apposed palms of the hands. Fire is produced by drilling it into the surface of another piece of the same wood quadrangular in section. Upon one surface of this a saucer-shaped hollow is made of diameter large enough to accommodate the end of the drill. On one side a V-shaped incision is made so as to cut into the saucer-shaped hollow.

To make fire the latter piece is placed horizontally on the ground with the saucer-shaped hollow uppermost, preferably with a flat arrowhead underneath it, and held firmly in position by the feet of the squatting Indian. The drill, held vertically between the two hands, is pressed firmly down into the saucer-shaped hollow and rapidly rotated, the downward pressure being maintained the while. The dust produced by the friction falls down the vertical V-shaped groove and accumulates in a heap on the arrow-blade. This dust soon shows signs of rise of temperature by changing colour, becoming first brown then black; and finally a little column of smoke rises from it. The glowing mass is then placed among a little dry grass or kaliete fibre, blown upon gently or waved through the air, and the grass bursts into flame. Not only is obtaining fire by this method very speedy, even a comparative novice like myself taking less than half a minute, but it is little affected by rain, the first few turns of the drill sufficing to dry the apposed surfaces of the wood.

An Indian Day

The toilsome day of the Natokoi Indian begins at or even before the first signs of dawn. Singly or in small parties the men depart for one or other of their favourite hunting grounds—the jungle that fringes the river or swamp, the open palmar, or the islands of forest (monte duro). Gliding stealthily along with the Indians in absolute silence it did not take me long to appreciate the handicap to success in hunting imposed by the wearing of European boots! The senses of hearing and sight are kept attuned to the highest pitch of acuity. A rustle of vegetation, the crackling of a twig, the note of alarm of a charáta or chajá, calls an immediate halt. The Indian knows that if he remains absolutely without movement any wild creatures will pass him unheeded. When on the move his sense of sight is concentrated mainly on the ground. Not only is he able to distinguish the trails of local animals with extraordinary skill but he forms a rough judgment of the time that has elapsed since the trail was formed, from the extent to which the displaced vegetation has recovered its natural position.

Along the margins of swamp and river the chief game animals are the large marsh deer or Cierbo (*Cariacus paludosus*), the Carpincho (*Hydrochoerus capybara*) and the coypu or nutria.

In the palmar the chief animals are the Pampa Deer (*Cariacus campestris*), the Ostrich or Ñandu (*Rhea americana*) and the wild Guinea-pig (*Cavia aperea*) and with the two last mentioned special methods of hunting are used. In hunting the *Rhea* the Indian conceals himself in an extinguisher-like covering of twigs and leaves and so camouflaged stealthily approaches the grazing *Rhea*. Periodically the latter raises his head to look around for potential enemies. The Indian then stands stock still without movement until the bird's attention is again concentrated on his grazing. Then the approach is renewed, stage by stage, until within range of arrow shot.

The wild guinea-pig or cavy, extremely abundant in the Chaco, spends most of its life in the obscurity of its tunnel-like runs in the dense grass. To hunt these cavies the Indians fire the grass in a wide circle, keeping open here and there gaps in the ring of flame through which the animals bolt. The Indian standing by has his bow ready bent and one sees some very pretty shooting as one after another the guinea-pigs are transfixed in spite of their small size and great rapidity of movement. Usually the guinea-pig is prepared for cooking by plucking out the hair, but occasionally the skin is pulled off whole and then it forms a useful bag for carrying honey or water.

In the monte the chief game animals are the charming little forest Deer (*Coassus simplicicornis*) and the two species of Peccary. The former is a beautiful example of camouflage by counter-shading, there being the most perfect gradation between the darker grey colour above and the yellowish white below so as to mask completely the relief of the body form and make it appear absolutely flat. Standing motionless within the margin of the monte the deer remains undetected until one's approach to within a few feet causes it to bound suddenly into visibility and flee.

The peccaries (which as already mentioned have come to resemble superficially the wild pigs of the Old World, although not closely related to them) provide palatable meat so long as one takes care to excise from the middle of the back a gland which produces an unpleasant secretion.

Great tales are told of the danger of hunting the larger white-lipped Peccary (*Dicotyles labiatus*) which goes in troops of as many as sometimes a hundred individuals. Hunters have been described as being cut to pieces by the sharp tusks of a herd which they have encountered, or as being treed amid a circle of peccaries waiting patiently and looking upwards expectantly at their wished-for prey. In point of fact there is no particular danger if they are hunted by the Indian method. Skirting along the edge of the forest one's first intimation of the presence of a herd of peccaries is the noise they make as they dig up the roots on which they feed. In this event we would take to the forest and make our way in silence to a point close to where the peccaries were feeding. Then a wild dash out with loud shouts into the midst of the affrighted animals, scattering in all directions, the Indians pursuing preferably young individuals with their clubs while I picked off more distant ones with the rifle. While we gave in this way no opportunity for the peccaries to recover their nerve we really had no opportunity of testing how they would behave if not upset by our sudden rush.

It may be that one of the Indians setting out from the camp goes in search of honey. Of this there were the three common varieties mentioned in the preceding chapter. By far the most important was the Katik or Lecheguana. The nest is a spherical papery structure a foot or more in diameter, containing in its interior concentric shells of the same material carrying the honeycomb on their inner surfaces. The honey was pleasant to eat—very pure sugar with much less pronounced flavour than our European bee-honey, while admixed with it were large grubs, probably of a beetle intruder—cool and juicy, going off with a plop in one's mouth like a ripe gooseberry.

The skill of the Natokoi in following ordinary trails was most impressive but still more so was their ability to track down the Katik wasp to its nest. Specially expert was my friend Chinkalrdyé (Plate XV). As we walked along he would suddenly point in the air and say 'Katik'. I would look in the direction in which he pointed and would see nothing, or at most some unidentifiable fly-like insect. Chinkalrdyé, however, had at once spotted it as a Katik wasp and—far more impressive—was able to distinguish whether it was unladen, on its outward journey, in which case he paid no further attention to it, or laden, on its homeward journey, in which case he marked exactly the direction of its flight, followed in the same direction, keeping a careful look out for other returning members of its community, and invariably succeeded in finding the nest among the branches of a tree.

He would now get out his fire-sticks and make a fire to windward of the nest, at such a distance that the stream of smoke was carried to the nest by the wind. In a very short time the wasps would begin to emerge and retire to another tree some little distance off. When the nest seemed to be entirely deserted Chinkalrdyé would climb up the tree, cut through the twigs to which the nest was attached, and bring it down in triumph. After we had made an adequate meal of its contents the remainder would be stowed into one or more guinea-pig skins and carried back to the camp.

Other Indians, though not Chinkalrdyé, when hunting Katik would envelop themselves as completely as possible in their garment when climbing the tree as a partial protection from possible defenders of the nest, which were armed with stings, though not formidable ones.

The other honey-collecting wasp, 'Bala' of the Argentines, is smaller in size, black marked with yellow and its nest of similar paper-like material to that of the Lecheguana but much smaller, somewhat pyriform in shape and provided with numerous tiers of cells arranged horizontally one over the other.

While most of the men were away hunting some would remain behind in the camp as a guard. During the heat of the day they would take to the water if available. They were excellent swimmers, the children learning to swim almost before they can walk. They swam with a side stroke and did not raise the hand above the water's surface. When conveying small articles across a river they held it above the surface with one hand and swam with the other. They swam well under water. After bathing they did not rub themselves dry; their long hair they dried by flicking the moisture out of it with the edge of the hand.

The women bathed too but apart from the men.

Curiously the Pilcomayo Tobas had no canoes and seemed entirely ignorant of the art of navigation. The paddle-like form of the implement carried by the women (p. 130) suggests at least the possibility that this may not always have been the case.

As the day wore on the hunters would drift back to the camp with their spoils. Food was cooked by the primitive method of broiling in contact with the glowing embers. Culinary evolution had not yet reached the stage of boiling or stewing. Indeed the rude earthenware pottery, dried in the sun, was not capable of withstanding the heat of a fire. Of vegetable food there would be the heart of the young palm (Chaik-Kūm) eaten raw.

During what remained of the day the Indians would sit about resting and doing odd jobs, especially attending to their weapons; shaping a new bow or making arrows. Such work took time as tools were confined to the flat iron arrow heads already mentioned, detached from their shaft and used as primitive knives.

In this region of the Chaco with its complete absence of hard rock we may take it that there was probably never a purely Stone Age comparable with that of other parts of the world. But yet there were in past times stone axes and hammers. We know this through the present-day archaeological collectors of the Chaco—the ostriches which, in the absence of pebbles, eagerly snap up any stone implement they may encounter. In Plate XVI (*b*) are shown two stone axe-heads and a hammer, each composed of volcanic rock from some far-distant region. To what extent they owe their presence in the particular locality where found to barter from tribe to tribe in former days or on the other hand to the wanderings of the ostriches from which they were obtained is a question which cannot be settled.

The first metal axes of the Indians were of the small tomahawk type shown in Plate VII (*b*)—a metal blade inserted in the head of a wooden club. No doubt its stone forerunner was similarly countersunk in the head of a club to give the necessary weight. But already at the time of my sojourn full-sized modern axes were beginning to make their way through the Chaco by barter from tribe to tribe.

While the men are away on their day's hunting the women remain about the camp attending to the young children and performing their own special communal duties.

Amongst these is the collection of Chaik-Kūm, the heart of the young palm, which provides the one normal vegetable element in the Indian's diet. To extract this use is made of the Wood Age representative of the axe of later stages of evolution (Plate XVI (*a*)). This consists of an enormous paddle-like implement of heavy cascarandá wood, which is held vertically and brought down heavily on the head of the young palm, the blunt tip of the paddle cutting its way down among the leaf bases and exposing the young and soft inner ones which alone are eaten and to a European are quite as palatable as raw turnip. It is interesting to contrast the potency of this implement actuated mainly by the force of gravity with that of its successor in the Stone Age and the age of metal which gains its effectiveness from the momentum given by the swing.

During its season the fruit of the Algarobo or Locust Bean (*Prosopis juliflora*) is in great request and some of the women set out from the encampment each morning to collect the 'Kamūp' as it is called—long slender pods whose soft substance contains much sugar and is alike refreshing and satisfying. The pods when brought into the camp are pounded in a rude mortar to separate the hard seeds from the surrounding matrix and convert the latter into a fine meal. This may be eaten as it is or mixed with water to form a drink, or kneaded up into a paste. In the early part of the season the fruit of the Mistol (*Zizyphus mistol*) is sometimes added to the Algarobo paste which dried by the fire forms a palatable kind of cake (patai).

The work of spinning and weaving is done by the women. The indigenous fibre—extremely tough, of considerable length and highly resistant to the effects of damp—is obtained from the leaves of the pine-apple-like plant *Bromelia pinguin* (Natokoi—kaliete, Guaraní—ivira, Argentine—chaguar) by simply scraping the leaf with the edge of a bit of shell. It is kept in the form of plaited skeins. It is twisted into thread and cord by the women who roll it with the palm of the hand on the front of the thigh.

The warlike Natokoi supplement their supplies of kaliete by raiding the Orejudo Indians who possess flocks of sheep. The wool is spun into yarn with the help of a spindle consisting of a slender piece of wood about 9 inches or more in length and inserted at one end into a heavy ball of hardwood. The yarn, as spun, is wound round the axis of the spindle close to the wooden ball, the free end being looped round its upper extremity to keep it from slipping. The spindle is twirled rapidly between the finger and thumb and when let go continues to rotate for some time on the end of the yarn which it keeps twisting while fresh wool is constantly being paid into its upper end. The spindles, by the way, are obtained along with the wool from the Orejudos and are therefore not to be included in the list of indigenous implements of the Natokoi.

The yarn, whether of kaliete or wool, is woven into fabric to form the garment worn by the men, usually dyed with vegetable dyes to form a striped pattern of broad, or alternating broad and narrow, stripes of dark brown or red on a ground of the natural colour, the fashion differing in different tribes.

The process of weaving I was never able to see on the Pilcomayo, but as it like that of spinning was taken from the Orejudo natives to the northward it will suffice to describe it as observed among them some years later.

The loom consists of two horizontal beams some 6 or 7 feet long supported by two vertical columns forked at their upper ends, and about 4½ feet in height. The yarn is looped vertically round the two horizontal bars to form the warp, red, brown or black yarn being used to produce the longitudinal coloured stripes. The horizontal strands (weft) are passed alternately in front of and behind the warp threads. To facilitate the latter operation alternate threads are pulled towards the weaver by loops passed round them and then held in position by a thin cane. This cane, passing as it does alternately in front of and behind the warp threads, serves as a guide along which the weaver passes her ball of yarn, no shuttle being used. The weft thread having been passed through the whole width of the web is then beaten down close to its neighbour and the process repeated over and over again until the whole length of the warp is filled up.

Towards sunset—at almost exactly the same time every day—the women set about their last communal duty, going off with an armed escort for a supply of water to last over the night. On these occasions they invariably travel at a characteristic swinging trot which they keep up the whole way. The water they carry in an earthenware water-bottle slung over the forehead by a cord.

As darkness falls the fire is carefully damped down so as not to betray by its flicker the exact location of the camp. It is only when the party feels assured of the absence of enemies in the neighbourhood that this precaution is omitted. Often in civilised life I have wondered whether the pleasure felt in sitting by a flickering fire, or listening to the ripple of a stream, may not be an ancestral feeling that has persisted through the ages from the times when our forebears were still at the stage of communal evolution in which these Indian tribes still linger.

As I became more intimate with the Tobas I found them much more loquacious than they had seemed at first. In the presence of strangers they are indeed silent and will sit for hours without uttering a word. They watch all their doings but say nothing. But in the absence of strangers it is quite different. When we halted during the day conversation was always going on and at night they would go on talking sometimes for hours. But eventually conversation gradually dies down. One after another the Indian divests himself of his garment, spreads it on the ground and lies down on it to sleep—with no protection whatever against the blood-sucking insects of summer or the cold—often intense—with hoar frost in the hours just before dawn—of winter. He sleeps lightly, undisturbed by the normal sounds of animal life but instantly wide awake and alert on the alarm note of chajá or charáta or still more the sharp crack of a twig which betrays the immediate proximity of danger.

Another mistaken first impression agreed with that of D'Orbigny who wrote of the Tobas 'taciturnes autant que possible, ils ne rient presque jamais'. This again is true enough of the Indians in the presence of strangers but in their absence it is quite otherwise, and faces are frequently lighted up with a smile. They have too a considerable sense of humour. One of the Toba girls was a great chatterbox: her name was Cōché. One day I said it should be changed to Cōchŭk—the chattering Urraca Jay. She was most indignant but the Indians generally much appreciated the joke and it became quite a habit to tease her by calling her Cōchŭk. They would occasionally play practical jokes on myself when bathing, swimming unseen under the muddy water and trying to pull me under, or at other times making fearsome narrations in my hearing of what was to be my fate.

Dependent as the Natokoi are on hunting and fishing, the location of their camp is determined by the abundance of game or fish in the neighbourhood. This in turn is conditioned by various factors, above all by variations in rainfall. During the dry season the larger animals tend to concentrate in the neighbourhood of the rivers and the more permanent bits of swamp. Again the wide-spreading fires while causing at the moment great destruction of the smaller game such as guinea-pigs, serve later to attract deer through the rich young pasture which springs up after a shower of rain. The dry season has also great effect on fishery—the fish being first concentrated in the shrinking water and later completely destroyed except such as have special arrangements for surviving through drought.

When on the move the party of Indians form a characteristic sight. Travelling as always in 'Indian file', a habit adopted to facilitate progress through dense vegetation, first come the chiefs and the older men, then the younger men each carrying only his weapons. Behind them the women and children. The women are heavily laden. Slung on the back of each by a string passing over the forehead is a large skin bag containing various stores; on the top of this rests a roll of dried cierbo-hide, or the large mat formed of giant rush-stems fastened together side by side which when unrolled can be used as a 'toldo' or shelter, several earthenware water-vessels of various sizes with their cord slings; on top of all frequently a little child astride of its mother's neck: the whole forming an enormous burden. Finally she carries in one hand the heavy 'paddle' of cascarandá often over 6 feet in length and weighing as much as seven pounds, normally functioning as an axe or spade but now used for steadying her footsteps.

Whenever possible the party skirts the edge of the hardwood forest, into which it slips silently on the appearance of danger: the women at once scattering through the interior of the wood while the men tuck up their garment tight round their waist so as to leave their legs quite free, or discard it in favour of the fringed girdle if they have one, test their bows to see that they are tightly strung and then take up their position just within the margin of the monte whence they can have a clear view of the approaching enemy without being themselves visible.

On one occasion when with a party of our soldiers following a rather doubtful party of Indians I made the fascinating discovery that the dark coppery red of the Indian's skin is a very effective camouflage, the Indians remaining undetected until we were almost on them. The obliterative effect of the colour is no doubt due to the fact that our perception of the dark shadows among green vegetation is, without our realising it, affected by its contrast with the bright chlorophyll green, tending to take on the complementary colour. On my mentioning this observation to that great naturalist Ray Lankester, he recalled having noticed in a particularly effective painting of woodland in a Royal Academy exhibition that the artist had introduced a slight reddish tint into the shadows. A similar effect in the reverse direction is seen in the momentary blaze of emerald green occasionally seen when the light of the setting sun is suddenly extinguished as it passes behind a highland mountain.*

When the neighbourhood of the projected new camp is reached, its exact location is determined, mainly by three conditions. It should be close to the edge of the monte, it should have drinking water close by, and it must not be in a depression in which water will collect in the event of rain. The spot being chosen, the women deposit their impedimenta. A fire is lighted, usually by a live ember brought from the preceding camping spot. In dry weather this can be done and obviates the use of the fire-drill. If a heavy storm threatens, a rude shelter may be made by bending branches of a bush over the sleeping spot: or one of the large rush mats (niyik) may be strung up to a bush or tree at each end so that one can find a little protection by crouching underneath or in its lee.

These Natokoi of the lower Pilcomayo had no more permanent shelters: no huts or tents: but it was interesting to find that the Indians who later on accompanied me to Paraguay, on their first sight of a civilised house called it by the same name as the toldo or shelter made of rushes—niyik pok—big mat—recognising as the biologist would say the analogy i.e. similarity in function of the two structures so unlike in appearance.

Living among the Natokoi made one realise how foolish it is to label such people as an inferior race. They are without the literary education which we are apt to regard as all-important. They are without books in which the advances in human knowledge are stored up and passed on to subsequent generations. They are without training in mathematics, that all-important method of technique. Pictorial art has not yet developed. Their religion is merely the natural development of fear of the unknown and its personification in the conception of evil spirits.

But on the other hand they are superior to ourselves in physical health—being kept at the highest level by the grip of natural selection and its remorseless elimination of the unfit. They are far ahead of us in their skill as observers, in their constant mental

* See *The Times* 23rd June, 1933.

alertness, and in their power of rapid and accurate interpretation of what they observe. Their code of ethics is of a high order —patriotism and loyalty to the tribe, love of family, self-sacrifice, honesty, kindness. All these taught a worthy lesson to the stranger who had come amongst them.

They and I taught each other much else: they the natural history of their environment, I something of the regions of natural history outside their own experience. On the *Bolivia* we had a small astronomical telescope and I taught them to look through this 'big eye' as they called it and study the surface of the moon, with its craters, the planet Venus with its phases like the moon, Jupiter with the four of its moons that were visible. Such things naturally led one to talk of other things new to them—the ocean towards which the great rivers ceaselessly flowed, the regions of the world where the hoar frost which they experienced in winter mornings was replaced by an all-concealing mantle of snow and ice.

Their respect for my knowledge of such things as had lain outside their ken was on occasion deepened by some mere accident. One bright sunny day in the early afternoon I happened by chance to catch sight of the planet Venus and pointed it out to my companion. It spread like wild fire through the tribe that I was able to make the stars visible by day!

To use the language of to-day it might be said that the education of the Natokoi was concentrated on citizenship, on the things calculated to make him when grown up an effective member of the community. During the period of infancy and early childhood the educative process is carried out entirely by the mother. During the next stage it is the other boys who fill the role of teachers. The small boy plays with them and his toys are a small bow and arrow fashioned by his father. Copying his older comrades he learns to use his bow and arrow at fixed marks. Then venturing further he pursues birds and other small game and now he gradually learns something of greater moment than ability as a marksman: he acquires skill as an observer. Still more important he acquires skill in interpreting what he observes. He learns the meaning to be read into the impressions however faint that he receives through his senses of sight or sound or smell. In other words during all this play, in which he naturally competes with his comrades, ever anxious to surpass them, he achieves not merely bodily skill but also mental ability. And naturally too the constant activity involved conduces to perfect physical health and strength.

The period of adolescence brings a further stage of education, for now he parts company with the mother and other women and begins to associate with the men. Accompanying them and watching their ways he acquires proficiency as a hunter of the animals used as food. In camp he acquires from them knowledge of his environment and of such of the plants and animals that inhabit it as are of importance. Religion, too, for he is told of the unseen evil spirits which are the cause of disease and other misfortunes. When all goes well the Natokoi, like many men of other races, hear no particular call to religion.

In this period he learns also the rigid code of tribal custom, infractions of which bring condign punishment. This code involves as its very foundation loyalty to the tribe. Within the tribe there is truthfulness and honesty but these virtues play no part outside its boundaries.

The tribal unit among the Natokoi I found much less definite than I expected. Each tribe is known as the men of so and so, e.g. 'Yordaik lyale' the men of Yordaik. The chief is simply the big man among smaller men. The tribe gathers round the man with outstanding qualifications for leadership—not the man of special skill in hunting or in single combat but rather the man of sound judgment and reliability. It follows that different tribes differ greatly in size—up to say 200: the able man accumulates round him a tribe proportional to his attractiveness as a leader: as this attractiveness diminishes with age so also does the number of his followers as they drift away to some other community. Chieftainship then among these Natokoi was an indefinite thing. There was of course no question of its being inherited and it involved no formal or official power. I would tell Yordaik I wanted something done. He would tell his men sitting round to do what was wanted: they would show no symptoms of having heard him: then he would get up and do it himself.

On occasion, for war against a common enemy, a number of neighbouring tribes would unite. In face of prospective attack by the Nimká the men of Yordaik were joined by those of Lyániteroi, Taashé, Whynache and Yokoidyi.

The Indians are slaves to etiquette. Their social manners are precise. The time to pay a call is the late afternoon just before sundown. The party of visitors is seen approaching in single file and the men come in first. They are invited to sit down and do so in a circle, the visiting chief beside the host. For a long time not a word is spoken. The visitors attend to their toilet. They comb their hair, paint their faces, and arrange their head-dress. When all is well, conversation begins but at first purely formal and stilted.

'We have come from the West.' 'You have come from the West.' 'There was much water en route.' 'There was much water en route.' 'We fought so and so's men.' 'You fought so and so's men.' 'We killed ten of them.' 'You killed ten of them.'

And so on, the host politely repeating verbatim each announcement of his guest—and then after a pause conversation becomes freer and more general.

One time I proceeded with Yordaik to pay a call upon another chief, Yokoidyi, and his men. On our arrival at their camp they spread skins on the ground for me to sit upon. The two Chiefs squatted down beside me and the tribesmen in a circle. Then an immense gourd holding at least a gallon and a half of 'luktagá' was placed on the ground in front of Yokoidyi, who handed the drink in a small gourd, first to me and then to the others round the circle—clockwise by the way, although in the southern hemisphere—suggesting that there may be something very old and ancestral in our custom of passing the wine clockwise at a civilised party. The luktagá, bright yellow in colour, sourish and very refreshing to the taste, had been made from honey and pollen and was very strong—'chim' as they called it. Although there was a huge quantity of luktagá in the gourd, it had to be replenished in order to go round completely, each man consuming about a pint—sufficient to make them all very talkative—denouncing in particular my enemies the Nimka and pledging their intentions of utterly annihilating them.

The luktagá is not always made from honey. During the Algarobo season it is made from the locust pods. These are placed in a large gourd, a handful taken out, chewed thoroughly and then put back, until the whole has been subjected to this process of communal chewing. To the stranger it is at first a somewhat trying experience, though the Indians have naturally perfect teeth, living as they do in conditions in which only the absolutely fit survive.

At such visits from strangers the women visitors came in some time after the men, one by one, their hostesses running out to meet them and relieve them of their burdens. They then disappeared from view, the women remaining apart on such ceremonial occasions as indeed they do during the greater part of their ordinary lives except when the tribe is on the move. In the early days of our intercourse with the Indians while we were watching one another's ways the women would always remain grouped together in the background.

It will be understood that disease as prevalent in civilised communities hardly exists among these primitive peoples, subject to the merciless supervision of Natural Selection. A comparatively trivial disability whether constitutional or due to accident is sufficient to render the individual unfit to survive. A mere sprained ankle or a simple Collis fracture of the wrist means inability to carry out the ordinary activities of hunting and fishing, and the wandering party of Indians cannot be burdened by invalids. And consequently one misses among them the feeble and crippled—so conspicuous a feature of the crowded life of our civilised communities. Not only is the disabled individual promptly eliminated, but even the tendency towards disability is being constantly weeded out so that, as in the case of wild animals, the race is maintained at the highest pitch of health.

An important factor in this health is the immunity developed in the course of generations towards the prevalent microbes of the locality. There is always however the possibility of an outbreak of disease being caused by an invasion by some alien microbe coming from some other species of animal or from some other human community. Dreadful ravages among the Natokoi were caused by the introduction of the spirochaete of syphilis by the Cristianos, who themselves had developed the disease in a milder form, due to its gradual attenuation in the Old World through the course of centuries.

The mode of life of the Natokoi in small isolated communities fortunately provides a potent safeguard against such invasions of alien disease.

It should be mentioned that the method of incantation is not restricted to the prevention and cure of disease, but is used to scare away the powers of evil in other circumstances, as for example before some prospective undertaking involving danger.

When for example we were about to start upon the expedition through the territory of the Mushcui, described in the next chapter, the night before our departure was marked by continuous incantation from sundown to sunrise to safeguard us on our way.

In their general sex relations the Natokoi were healthily normal: the young women attractive, evasive and modest; their often beautiful and graceful figures little concealed by their short garment and even this they readily discarded without any embarrassment. In a sudden plump of rain it would be whisked off and rolled round the head to keep the hair dry.

Sometimes, as I luxuriated in the coolness of a deep pool, a couple of them would join me, playing about in the water like a couple of naiads, the air echoing to their merry laughter or to the soft alluring crooning, which is one of their arts in the enmeshment of the other sex.

There was no ceremonial marriage, but couples seemed pretty stable in their attachments in spite of the very general polygamy—the original partnership being added to from time to time by a younger partner, until the number reached might be five or six.

In my diary (p. 147) I mention how moral lapses on the part of attached women are dealt with.

They reproduce slowly, and a factor no doubt involved in this was the prolonged period during which the mother nursed her infant.

Apart from burying a new infant with its dead mother there was no practice of infanticide. So I was assured by the Natokoi, though they knew of the practice among the Orejudos.

Philosophy. In the philosophy of these primitive men there had not yet come about the sub-division familiar in later stages of evolution—into science, medicine, religion. Science, the most ancient part of all philosophy, was still predominant—the lore of the environment in which they lived and upon a knowledge of which their lives were dependent. When actual knowledge did not suffice to explain particular phenomena, at least if unpleasant in their nature—for things that were agreeable seemed not to demand explanation—they drew upon the existence of supernatural beings and thus constituted religion of a primitive type. Where the particular phenomenon was that of disease, the dealing with the evil powers concerned constituted the beginnings of medical science.

The method of dealing with such powers of evil, whether concerned with the bringing about of disease or any other misfortune, was always the same—incantation. In the case of illness one of the older men acting as physician bends over the patient and chants at the part afflicted. Night is the time these spirits are abroad and the incantation will go on all night through. In one case where a woman was suddenly taken ill it seemed clear that an evil spirit had got in amongst them all, and the whole of the men joined in the chant at the top of their voices.

The only kind of direct medical treatment that I saw was a kind of blood-letting in which they had great faith as a cure for local trouble. With a knife, i.e. one of the flat iron arrowheads, a number of light cuts are made through the skin of the part affected and then sucked.

However, during my stay they developed great faith in the potency of my drugs, although when expressing their high opinion of them, they assured me that they had not lost faith in their own system of medicine of which they would give me the benefit were I in trouble: a contingency which happily did not arise.

Art. While able to mould clay or wax into crude images of natural objects such as the body of man or beast, the Indians showed at first no understanding of pictorial art—the conventional method of representing solid objects on a flat surface. Pictures even of familiar things like a man or a jaguar or a parrot conveyed nothing to them.

However, when once the meaning of pictures had been explained they very quickly grasped the idea and very soon they were able to interpret pictures. It then became for them an enthralling experience to sit round and watch as I turned over the pages of a book on natural history, excited exclamations 'ah ah' betraying their interest in recognising some familiar local animal or wonderment at some unknown type such as an elephant or a giraffe.

The climax of their artistic evolution came one day when I was making a drawing while the Indians sat around, and Cacique Lyaniteroi suddenly said that he too would make a picture.

I handed him a sheet of paper and a pen and told him to make a picture of his man Chineratalοi.

A facsimile of his picture is here reproduced, and is of importance to the student of the evolution of graphic art as being the first picture ever made by a particular race of men. Two points of detail that are of interest are (1) the accuracy in the number of fingers and toes, and (2) the relative unimportance—so far as expressed by size in the drawing—of head as compared with hands and feet!

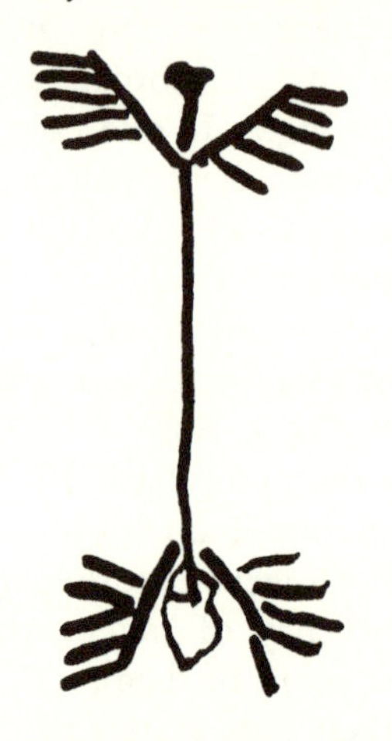

While pictorial art is seen here in its first beginnings, decorative art on the other hand has reached a distinctly higher level, as appears in the patterns of garments and bags, or by those incised on the surface of gourds.

Measurement. The Natokoi had no definite measuring instruments, though they would indicate size with their hands in describing a small object or short distance. For geographical distances the unit was the day's journey—necessarily a very variable quantity. The use of any such unit was restricted by the limitations in the power of counting. There were no words for numbers—merely the holding up of the fingers of the left hand up to the limit required—first the thumb then the other fingers as required in succession. Five very commonly was the limit: one specially mathematically minded might reach seven—adding the thumb and forefinger of the right hand to the five fingers of the left. There was no use of ten as a unit—but an idea comparable with infinity—to express the trees in the forest or stars in the sky—was expressed by placing the hands down beside the feet, the twenty fingers and toes being beyond comprehension as a number.

Reckoning of Time. The Indians naturally had no clocks. They indicated a particular hour of the day by pointing at the position in the heavens that the sun would be in, indicating midday for example by pointing to the zenith. My watch hung normally in the deckhouse of the *Bolivia* and there it was an object of great interest to the Indians, inhabited as they believed by a spirit which caused the hour-hand to indicate midday in the normal fashion by pointing upwards to the zenith. Chinkalardye in particular would often ask permission to take the watch down from its peg so that he might listen to the voice of the little spirit inside. He would take it down, convey it carefully into the open air and would remain for a long time squatted down, the watch held by both hands first to one ear and then to the other, while he listened to the ticking voice of its inhabitant. There was naturally no conception of the hour or any other unit of time smaller than the day, and the use of this latter was restricted to five or occasionally six or seven by the limitations in counting already mentioned. The next unit of time above the day was the moon (alwoik) of which the Indians recognised the phases—crescent (nēnagoishagĭm alwoik), half (napēgēdadiōrnī alwoik), and full (gewōgōchĭgĭnĭ alwoik) but there was of course no recognition of a definite number of days in the lunar period.

The year was marked by the season of the particular fruits, the onset of wet and dry seasons being very irregular. Here again there was no recognition of the recurrence of annual phenomena being separated by a definite number of moons.

The limitation of the capacity to count involved necessarily the absence of data as to the age of the Indians beyond early childhood. There were men and especially women who to all appearance were quite old, but I could not convince myself that life in their strenuous conditions extended beyond what we should call early middle age.

Right-handedness. In view of the relatively primitive stage of the Natokoi it is worthy of note that they are right-handed to a similar degree as are Europeans. This is in agreement with the view which I hold that the evolution of right-handedness in man took place in intimate association with the evolution of speech. The faculty of speech is known to be linked up with the left cerebral hemisphere. But it is this same hemisphere which dominates all muscular movements of the right side of the body apart from the head region: if the left hemisphere of the brain is injured the resulting paralysis is on the right side of the body.

Now it may safely be assumed that the gradual evolutionary advance of the function of speech in the left hemisphere would be accompanied by a corresponding advance in the perfection of its blood circulation, and this improvement in its blood supply would give the left hemisphere an advantage over the right in carrying out its functions other than that of speech. In other words the muscular arrangements on the right side of the body would have a certain advantage over those of the left side so far as concerns the brain centres dominating them.

Now one of the most striking features in the evolution of man has been the development of skill in the use of the fingers. In the development of such skill during the life of the individual it will on the whole pay to concentrate training upon one of the two hands: it is only in exceptional cases that so-called ambidexterity is more valuable.

It would seem that the relations with the brain that I have indicated are sufficient to give the right hand the slight advantage over the left which has led to man becoming right-handed.

In connexion with this interesting problem in human evolution I may quote an experience of many years subsequent to my stay among the Natokoi. I was looking through the portion of the official history of the 1914–18 War dealing with head injuries. Case after case in which the faculty of speech had been destroyed turned out to have an injury on the left side of the head. Then an exception came to prove the rule: a case of aphasia in which the head injury was on the right side. This at first sight startling contradiction of the general mass of evidence turned out, however, to be quite the contrary, for the record showed this particular individual was left-handed!—in other words here too the more highly skilled hand was linked up with the side of the brain concerned with the faculty of speech.

Gambling. In spite of the fewness of their possessions, the Natokoi were inveterate gamblers. They had two games of chance; the one played with a couple of cubical dice of bone the faces marked by diagonal lines, the other with four short pieces of split cane placed crosswise on the back of one hand and thrown into the air on the chance of their falling to the ground crosswise. The score was kept by sticking an arrow obliquely into the ground on a scale marked by transverse lines. In these games they would gamble away their dearest possessions, fire-sticks, iron arrow-head, even their bow.

Economics. The Natokoi had not yet reached the stage of having money or currency, that system of tokens which has played so great a part in the affairs of civilised nations as a means of short-circuiting the cumbrous operations of trade by barter. It follows that they were without the competitive accumulation of private monetary capital which has played so all-important a part in the development of civilisation. It does not follow that they were without personal capital but it was in the form of what has been called biological capital—human capacity and its resultant achievement. In the former comes firstly the inherited capital consisting of the individual's heritage of health and strength—physical, mental and moral—received through the parents and on the other hand the increment of this heritage during the life of the individual as a result of his education and practical experience. This latter type of biological capital the Indian possesses to the full, but yet it is in this that his great contrast with civilised man is apparent, for on the one hand his wandering mode of life prevents the accumulation of material property, while the absence of writing and printing prohibits the transmission of acquired learning from generation to generation, and through this the building up of the body of civilised knowledge.

It will have been gathered that the individual property of the Natokoi Indian is restricted to his wives, his weapons and a few odds and ends, such as his garment and his fire-sticks.

Of communal property there is nothing beyond the loyalty and helpfulness of fellow members of the tribe. The area of country occupied by the tribe has no precise limits: the encroachments of neighbouring tribes are held in check merely by the power to repel them.

Domesticated Animals. Apart from those which they obtain by raiding the Nimká or Mushcui the Natokoi had no domestic animals, with the one exception of the dog. The fact that the South American Indians already possessed dogs before the advent of the white man is testified to by the fact that the dog has commonly a purely Indian name while other domesticated animals are called by names more or less obviously corruptions of the Spanish name. The Natokoi are no exception to this rule. The dog bears the purely Indian name 'piōk': while the cow is called 'waiky' (Sp. vaca) and the mule 'muriká' (Sp. burrico).

The Indian dog does not bark—this characteristic, like that of the non-crying infants, being due doubtless to ruthless elimination of individuals showing any tendency in this direction.

The dogs hang about the tribal encampment and I did not realise that they had any attachment to a particular master until

I was enlightened by a personal experience. I happened to be prowling about one hot afternoon when I encountered a party of Indians enjoying their siesta. Reconnoitring carefully I saw they were friends of my own and I decided to give one of them a real scare. I slithered noiselessly towards him and was just on the point of grabbing him by the throat when my right side was seized by a pair of formidable jaws belonging to a large dog which without my knowledge had been noiselessly stalking me while I stalked his master. There was much merriment among the Indians over my misadventure.

Personal Relations. As will have been realised I had the advantage of gaining real friendship and affection from the Natokoi Indians. How real it was came out clearly from their voluntarily accompanying me without thought of pay or reward upon the journey to Asunción through the country of the hostile Mushcui.

During the early stages of our sojourn in the country of the Natokoi I had been subjected, all unknowingly, as already mentioned, to constant surveillance as I pursued my solitary hunting and collecting expeditions. And so they learned how closely akin my interests were to theirs. Later on, when I had been accepted as one of themselves, my relations with the Indians became indeed those of the closest friendship. But they automatically took a subordinate position. They would bring me all the tit-bits in the way of food: and whenever we stopped on a journey, bags of honey would be produced for me to select the kind I liked best. When camping at night they would take special care to make a comfortable bed for me while they lay simply on the ground without bothering to collect any soft grass and reeds for themselves as they did for me. These Indians did not wash: their skins remained as naturally clean as that of any wild mammal. But they perceived that I customarily performed a strange ritual, dipping my hands into a pool of water when we came to one and rubbing them together; and Yordaik the Chief promptly took upon himself to minister to this ritual—filling his mouth with water from a guinea-pig skin and squirting it on to my hands held out in front of me while he rubbed them between his.

Though they did not wash, the Indians loved to bathe in the hot weather, the sexes rigidly separate. They would lark about and enjoy grabbing one by the foot and dragging one under water. On the other hand they were entertained when I retired under the surface of muddy water and remained invisible as long as I could hold my breath, keeping myself down by clinging to roots. I had considerable lung capacity and the Indians were much intrigued when I managed to make a good deal of sound under water by forcing air out between my tightly apposed lips.

Occasionally the friendly sentiments of the Natokoi were a little embarrassing, as when Yordaik represented to me the wishes of certain young persons that I should settle down definitely with them. The position was trying, but in spite of the intimacy and friendship of my relations with the Natokoi I managed to avoid closing the gap which naturally separates the different races of men and which is a potent safeguard in their dealings with one another.

CHAPTER VIII

THE HOMEWARD JOURNEY

1st December: Difficulties about drinking water for men and mules —the wells being dried up—and lack of fuel conveniently available decided us to-day to make a start downstream in spite of the Fortin being not yet completed—only one side of the stockade being in position. So steam was got up, the dam broken through, and we moved down with the flood carrying with us a number of Indian passengers—Yordaik, Whynarché, Chigmaki and Yorichidyi—who appeared greatly to enjoy the thrill. We made quite good progress until about 3.20 p.m., when we were brought up by the breaking of the eccentric rod and one of the cranks of the starboard paddle wheel.

2nd December: Hunting with Chimaki. Striking across the open palmar, we reached the monte duro and skirted along its edge. Our first victims were a Charáta and a pair of Boie's Woodpeckers, the gorgeous scarlet crests of which Chimaki wanted to adorn his head-dress. We had gone about a quarter of a mile when Chimaki suddenly exclaimed *Yŏlŏ*, and pointed ahead where a couple of Collared Peccaries were to be seen feeding on a patch of burnt ground close by the edge of the wood. I made him put on my conspicuous pith helmet while I stalked stealthily up to within 70 yards or so before the peccaries caught sight of me and took to flight. The first one that I fired at got away, the bullet merely scraping his side, but my second barrel caught the other, smashing his neck, so that he dropped without a struggle just a few yards short of the shelter of the monte. We caught a glimpse of two youngsters following the Peccary which I had killed, so, bolting after them, we managed to get between one of them and the edge of the monte, whereupon it lay down and enabled us to seize hold of it. Covering up the Peccaries with grass to conceal them from the vultures, we made a diversion to track down some Lecheguana wasps and after a good feed on their Katik we returned to the Peccaries, cut up the big one into manageable bits, and duly reached the boat about 9.30.

By a little after breakfast time, the paddle wheel had been put in order and we were able to make good progress, reaching our old stopping place at Dam VI just before sundown. We had been taking on board en route large numbers of involuntary passengers —the red and black social spiders, in a great state of bustle fixing up their webs.

3rd December: Again hunting with Chimaki, this time away to the South-South-East, through open country, grassy palmar with an occasional clump of Tinte-tacos (*Prosopis adesmioides*) or a solitary Vinál, to the hardwood Monte beyond. As we skirted along the edge of the monte we heard the well-known voice of the Charáta in such volume as to tell us that many birds were taking part in the concert. Presently we came to an open glade leading into the forest, and penetrating this we soon came within sight of the vocalists in the topmost branches of tall Guayacan (*Caesalpinia melanocarpa*) and Molle (*Tecoma* sp.), trees which towered up above the general level of the Arrayans (*Eugenia uniflora*) etc. which formed the bulk of the monte. The Charátas did not see us and we crept stealthily along under cover of the bushes until we got opposite them and then plunged into the monte, Chimaki leading, threading his way with bare legs through the maze of Caraguatá and other prickly spiny plants, and beneath and between the numerous lianas, while I followed as best I could, getting many thrusts from the tips of the caraguatá leaves. In a short time, we were close to our prey, their loud and harsh concert gave place to their no less characteristic 'cheeps' of alarm as they paced and hopped about the branches watching the intruders below. One after another, a dozen fell to the ground, Chimaki plucking them and hanging them in his belt. Half a dozen Bluefronted Amazon Parrots completed our bag, though we saw fresh traces of Collared Peccary, Cierbo, and Jaguar.

4th December: We have been having a number of sham duels of late to study the Indians' fighting technique, but to-day we had a real one. I had arranged with Yordaik to go off on a hunting expedition, but at the time appointed he came running to tell me 'Cristianos peleando'—the Christians fighting—and sure enough, I found two of our men in mortal combat. Sergeant Gomez, it appeared, had given an order to Olivera, one of the troopers, which had not been obeyed with sufficient alacrity to satisfy the sergeant, who promptly seized his sabre and proceeded to inflict chastisement. Olivera, however, immediately drew his facon and attacked vigorously. The sergeant thereupon drew his knife also and the two went at it merrily—the sergeant having a weapon in each hand. The sergeant received a deep cut through his deltoid muscle just below his shoulder, while Olivera got three long cuts through the scalp, one on his back and one on the back of his left hand which nearly cut the hand in two, completely severing several of the tendons and four arteries. The sergeant was just on the point of finishing matters by running his opponent through, when Nelson dashed up and the drama concluded, just short of becoming a tragedy. Olivera had lost a lot of blood and kept me busy for an hour and a half, stitching him together with an ordinary needle and thread. The tendons never joined up, so his hand would remain clenched for life. The bits of scalp did join up, but the tufts of hair projecting in different directions reflected little credit upon my skill as a surgeon.

8th December: Hunting with Chineratoloi, Chigordyí and Yaraitlik, the Indians wearing their 'Kádŭpĕlŭh'—rough moccasins made of deerskin—to protect their feet from the harsh cutting grass and the sharp recurved spines of the palm leaves which litter the ground everywhere in the palmar. Cutting across the palmar to the monte, we skirted along its margin but saw nothing of interest except fresh tracks of Collared Peccaries, so we decided to go over to the big esteros which border the Brazo occidental, the other branch of the Pilcomayo, and which seemed a likely spot to find Cierbo. For quite a distance we skirted along the margin of the swamp, keeping a close lookout for anything suspicious amongst the tall rushes and grass. For a long time we saw nothing, but at last we heard a soft whistle and looking back, saw Chigordyí signalling that a Cierbo was in sight, proceeding in a direction opposite to that in which we were going. So we ran quietly back some distance and paused opposite a gap in the rushes across which the cierbo would probably pass. We had not long to wait, for presently a big antlered head appeared and I aimed at where I judged his shoulder ought to be. He dashed wildly off, but just as I was about to fire my second barrel he dropped dead—another example of the effect of a heart shot. We covered up the carcass to conceal it from vultures and walked on for a long distance without seeing any more game. It was only on retracing our steps to skin and cut up the cierbo already killed that Yaraitlik espied a Guazu vira (*Coassus simplicicornis*) walking along among the trees and brush. Chineratoloi and I hurried off to circumvent him and at length got quite close, Chineratoloi pointing excitedly. At first I saw nothing, but presently became conscious of a beautiful little head gazing at me motionless at a distance of about thirty-five yards. It was a perfect case of camouflage by countershading, but there was no time for prolonged admiration and a charge from my shotgun rolled him over. We soon had the cierbo skinned and cut up, some pieces being roasted for our present refreshment, and then we started on our homeward journey to the *Bolivia*—I with the Guazu vira across my shoulders, and the three Indians each with a good load of cierbo. It was a hot tramp through about four miles of palmar with no shade from the vertical sun, and I at least was pretty tired when we reached our camp.

That evening I witnessed a strange spectacle concerned with the disciplining of one of the women of the tribe detected *in flagrante delicto*. In such cases, justice is inflicted by a committee of her more virtuous sisters, who proceed to give the culprit a severe pummelling. On this occasion I was roused by a tremendous din, and on getting up to investigate found that the frail culprit had got sympathisers who took her part and now the whole body of tribal matrons were engaged in a ferocious battle, the two sides being practically equal in number—the opportunity being, I suspected, taken advantage of for the purpose of paying off old scores. The women were in two ranks facing one another, their eyes flashing and both fists clenched and stretched out towards the enemy. Every now and then one would make a dash with her fist at her opponent's face and a tremendous scrimmage ensued; nothing was heard for a few seconds but a succession of thumps. After a bit there would be a pause, the two lines falling back a few paces and breathing fire at one another until they set to again. During all this time, husbands and brothers stood around, lighting up the scene with torches of palm leaves, shouting out words of encouragement to the dusky combatants, and looking on with what seemed appreciative grins.

At length, however, matters became serious; one dame was nearly shoved into the fire and others began to use sticks and the formidable palm leaf stalks with their sharp recurved spines, and then Cacique Yordaik dashed in among the women, holding up his hands and ordering them to desist. Husbands helped to separate the combatants and presently all was quiet—though for some days unfriendly scowls and bruised and scratched faces interfered much with normal comeliness.

9th December: Late last night three Orejudo spies were detected by one of our sentries prowling about the neighbourhood of our camp, but nothing happened.

11th December: A great disappointment; illness preventing me from going off on an expedition to the Orejudo country to the north westward—with Candioti, Nelson, Jack, twelve soldiers and two of our Caciques—Yordaik and Lyaniteroi.

The party was away for a fortnight and raided a tolderia of the Orejudos, carrying off fifty sheep and some maize as well as various odds and ends left by the owners who had at once fled on their approach. A little later I had an unpleasant experience. I was busy collecting in an island of monte to the southward when I suddenly realised that the forest was on fire. Going in one direction and another I found progress barred by the flaming vegetation, but at last I found a space where the fire had burnt out and was able to dash through it, singed but safe, and got back to the ship, taking care to keep out of the way of possible ambuscades.

Returning to investigate with a party of Indians, we found fresh trails of men and mules and it became clear that our friends the Orejudos had had the idea of finally disposing of me by incineration.

1st January, 1891*:* A year to-day since the expedition left San Fernando. A commission has started for Puerto Pilcomayo—Sgt. Gomez, Zapata, Lopez and Diaz: with Higginson and a couple of spare mules.

4th January: The trouble between Nelson Page and Lt. Candioti, which has been brewing for some time, came to a head to-day and the latter departed for Puerto Pilcomayo in high dudgeon with a couple of men. Nelson told me alarming stories of Candioti's expressed intention of abandoning the ship and having Henderson and myself shot.

In the afternoon Lyaniteroi came with the news that the Orejudos were out to revenge themselves for the recent raid and there had already been a big fight. The Indians around us are off to join their comrades.

11th January: Yordaik and a few of his men returned to report that the Orejudos had actually been on the march to attack us here, but had been forced to turn back by the absence of fresh water.

15th January: A very hot day with slight north wind. The shade temperature reached 109° Fahr. and the surface layer of the river water, deep green through the presence of innumerable unicellular algae, was almost scalding, though it was delightfully cool underneath. In the afternoon an Indian arrived with the news of more fighting with the Orejudos in which two of our friends have been killed. He also announced that Whynarche with a large body of Indians was on his way to join us and would arrive in three days' time.

The Sergeant had produced written orders left by Candioti instructing him to demand from Nelson full army rations of beef —3 lbs. a day—for the soldiers—a preposterous idea when we have to do our best to economise the supply—the more so in view of the fact that our dried beef (charqui) has a nutritive value much greater, weight for weight, than fresh beef still containing its moisture. It is no doubt an attempt on the Lieutenant's part to revenge himself on Nelson.

16th January: Hunting in the morning, Chinkalrdye and I encountered about 20 Indian women with four men encamped by a water pool. The men told us that they were bringing down their women to be near the *Bolivia* in view of the approach of the hostile Orejudos.

In the monte near Dam III got half a dozen Charátas, two by a rather remarkable chance shot as they were crossing the line of fire in opposite directions—one from left to right, the other from right to left. In the monte were many Jays—both Azure and Caerulean. As usual when in company the latter even exceeded the former in bold inquisitiveness. They were always the first to catch sight of us and kept hopping about among the branches round us raising the alarm by their harsh *caa-caa-caa.*

I also got a new Cuckoo which the Indians call 'Wuchen' (*Diplopterus naevius*) and believe that possession of even a fragment of this bird brings the owner luck in their gambling game of dice.

18th January: In the early morning the Commission arrived from Puerto Pilcomayo, bringing news from the outer world and also incidentally orders to the soldiers that they were no longer to apply for beef but were to eat their mules, and that they were to refuse to do any work for the ship.

19th January: The *Bolivia* made a fresh start down stream, but at little more than a hundred yards from the starting point was again high and dry, the water having drained away owing clearly to the breakage of the next dam. This is the second time a dam has broken just as the ship got under way and we cannot help suspecting foul play on the part of the soldiers. In the morning a Commission departed for Puerto Pilcomayo taking with them all Candioti's belongings. The journey there by mule now takes 8 or 9 days instead of five, as the Paraguayan Government has forbidden trespass on their territory. After breakfast it turned out that Cabo Diaz had disappeared with his belongings so he would appear to have deserted—the last survivor of our original party of troops. As we were turning in Nelson came into our cabin looking rather upset and imparted the startling information that he had just overheard a conversation among the soldiers from which it appeared that they had a plot to desert in a body, after murdering those of us on board. This was unpleasant news and we held a long council of war as to what was to be done. The essential thing seemed to be to endeavour at once to get into touch with headquarters, and I volunteered to try to get through to the Paraguay if one of the Tobas would accompany me as a guide. So it was decided that I should try to get off the morning after next, ostensibly on one of my usual hunting expeditions. In the meantime we would keep a specially strict watch throughout the night and I would talk matters over with Chimaki. When I did so two things became clear (1) that the Tobas had already suspected that the soldiers were up to something, and (2) that their feelings towards the soldiers in spite of their superficial friendliness were actually of bitter hatred, intensified by the maltreatment of their women and their infection with virulent disease, and that they would relish nothing better than to have a final reckoning with the Cristianos.

When I told Chimaki of my projected attempt to get through to Paraguay he was most enthusiastic, said of course he would come with me and talked much of the fate that would befall any Mushcui we encountered by the way.

Next morning I decided to take Chinkalrdyé with me as well as Chimaki, and then Lyaniteroi came along and said he would come too. I was not altogether glad to have him as I felt less confidence in him than in the other two, but Yordaik praised his powers of pathfinding very much and he will no doubt be of use as long as he behaves himself. Of the two, Chimaki is one of the best hunters in the tribe, while Chinkalrdyé is unequalled as a collector of honey.

January 22nd: Rose between three and four o'clock and made final preparations. I collected together a scanty supply of provisions—a little flour, salt, charqui and biscuit, and packed them into a pair of saddle-bags. The Indians were all awake, sitting round their fire, while Lyaniteroi chanted a weird and mournful incantation—to keep all evil spirits from us during our journey. As day dawned the three Indians came over to the boat and Nelson and I went ashore to the camp and saw the four mules which were ready lassoed, but we also recognised that the soldiers had picked out particularly vicious animals—two of them being probably the worst in the whole troop. None of the men gave us any assistance in saddling up, and from the ugly looks directed at us I could not help wondering if they had some suspicion of our project, and I hoped that we would manage to get clear away without any actual trouble. With the exception of my rug, I took no impedimenta of any kind—knowing the frightful nature of the country through which we should have to pass: that is of course exclusive of weapons, of which I took my express—with 60 cartridges—and shotgun, which I gave into Chimaki's charge. As always I carried at my belt my ·476 Webley revolver and long knife. Lyaniteroi and Chinkalrdyé carried their bows and arrows and the latter in addition his tomahawk for use in collecting honey. Hung on to my saddle I had a small enamelled saucepan to serve for drinking and for more elaborate cooking than the simple asado (roast) should time afford. For the purpose of getting fire, all the Indians had their fire sticks.

We had a good deal of difficulty in saddling up the mule I was to ride, as it kicked and bucked and bit until at last we tied a jacket over its head, which kept it quiet until I was in the saddle. The Indians were to ride bareback, but two of the mules proved quite impossible to mount, so we decided to leave them behind and let Chinkalrdyé and Lyaniteroi follow Chimaki and myself on foot. When we were mounted, Nelson took a photograph from the upper deck. Old Henderson and he waved an adieu and wished us success, and we set off.

Chimaki looked admirable with his shining coppery skin, black hair and white ostrich plumes. He and I started off alone, the other two being told to follow on and overtake us. Making a circuit for some distance so as to avoid the soldiers who were guarding the live stock, we struck the trail a few miles down the river, and followed it at a walking pace so as to give the others a chance of overtaking us.

About eleven o'clock we heard the curious vague prolonged whistle, which swelling gently on the ear and dying away again as imperceptibly, tells of the proximity of young Ostriches, without betraying the direction in which they are. However, Chimaki was quite equal to the task in hand, so we dismounted and tied up our mules to palm trees. Then Chimaki cruised cautiously about, and at last the whistle stopped suddenly and a young Ostrich about 18 inches high started up. He went in pursuit and after an exciting chase, for it ran very rapidly and doubled hither and thither, he managed to kill it with a blow of a stick he had picked up. We also saw the old bird but he was too wary and was off like the wind before we could approach within shot. Pursuing our way, we had gone only a short distance when I happened to look back and found the Indians overtaking us, but not two only as I had expected but six. They were all in full paint and plumes and fully armed: they looked very fine, wending their way between the trees. As I could not understand the change in numbers of my party, I pulled up my mule and waited for them. Lyaniteroi explained that he had consulted with the other Indians and that, owing to our having to traverse the territory of the hostile Mushcui, they had decided that it would be wise to have a larger number than I had originally proposed. This explanation was quite satisfactory and I was well pleased to notice that the additional five men were all special friends of my own. The complete list was Lyaniteroi—cacique, Chimaki, Chinkalrdyé, Sautranukdyî, Chigmaki, Yorichidyî and Chēnrai. Of these Sautranukdyî was an older man with only one—extraordinarily efficient—eye. He always walked with head cast down, peering at the ground in front of him and carrying on a running commentary on all the happenings disclosed by the appearance of the ground and its vegetation—what animals had passed and roughly how long ago. He carried only a club and had come with us as being considered the most skilled of all the tribe as guide and 'pathfinder' and as such he proved himself of immense service. All the other Indians were powerful, able-bodied men—the pick of Yordaik's men in fact—and I could not help feeling proud of my little army.

Pretty early in the afternoon we reached the fort—Fortin Nueve, where we left the men of the 9th regiment on April 1st of last year. The 'fort' consisted of two or three rude huts built of Carandai stems and thatched with grass. As the weather looked threatening, we decided to remain here for the night. We took up our quarters in the largest hut and soon had a fire blazing just within its shelter. Some of the Indians went out to collect a sufficient supply of fuel for the night. The others watched me as I cut up the charqui I had brought with me, and mixed it with the flour and broken-up biscuits. With a little water from a pool close by I made a stew, which with the little ostrich was just sufficient to give us all a good feed. In the evening it rained heavily. I turned in almost immediately after dark upon a heap of grass which the Indians had collected for me in one corner of the hut.

Round the fire the Tobas sat conversing in low guttural tones about our expedition and about their foes the Mushcui. Lyaniteroi went out into the darkness, and we heard him utter a short soliloquy and then chant one of his long strains of the usual weird and wild character—its burden being in low bass notes, scarcely audible—with a frequently recurring high-pitched refrain uttered with all the strength of his lungs. After he finished and lay down by the fire the voices of the Tobas gradually died away as they fell one by one asleep and all was now silent but for the voices of nature around us: the soft two syllabled cry of the Wuchin, and the harsh *beh-beh-beh* of the Pōtrunrut stood out as it were in relief from the soft continuous concert of crickets and frogs. And these too gradually faded away into the silence of sleep.

23rd January: At the first signs of dawn we made a start. The two mules gave us comparatively little trouble. We still continued to follow the track. Every now and then we came across the trail of a man with sandals—obviously not more than a few days old—and no doubt made by Corporal Diaz. By the edge of a swamp I shot a fine fat Cierbo and as there was fresh water close by we stopped here for breakfast. A mile or two further on we crossed over to the north bank of the river—the water not being more than 18 inches deep and of course still intensely salt and quite undrinkable. Lyaniteroi I find is of no use whatever in spotting game. He usually rides Chimaki's mule. His head is bent down and he appears buried in profound meditation. To-day as we were traversing a wide expanse of open country bounded only by a low line of palms on the horizon, he pulled up his mule suddenly—apparently overcome by the immensity of things—gazed into the distance with an expression of extreme solemnity for some minutes, then passed his hand slowly and dramatically round the horizon—without uttering a sound—and then passed on. At other times a sudden inspiration would seize him and he would deliver a long soliloquy in a deep and low tone of voice, mostly on the beauty of the scene as a hunting ground—all the others meanwhile preserving a respectful silence.

We had a long and wearying day's march, the heat being intense and not a particle of shade to be found anywhere. Just as it was falling dark we reached a large monte and I decided to camp by its margin, in spite of protests from the Indians, the meaning of which I did not at the moment grasp. The sky was now dark and threatening and we saw that the intense heat of the afternoon was the presager of a tempest. We tethered the mules close by as a precaution against tigers. Yorichidyî had meanwhile lighted a fire and was accumulating a good supply of dry logs for the night. I had shot a fine muscovy duck in the morning and this, along with the remains of the deer, was to serve for our supper. We sat close in round the fire although it was intensely hot and close, as the mosquitoes were partially kept off by the smoke, and it was cheerier too, for the night had closed in black as pitch all round, fitfully illumined by an occasional dazzling lightning flash. The logs crackled merrily. In the distance we heard the low rumble of the thunder gradually approaching. Some way off we heard the roar of a tiger and in the monte close by the deep *boo-hoo-hoo* of a Ñacurutú. The duck began to give out a most appetising odour, and weary and tired as we were we looked forward much to the delights of a good meal. Then a heavy raindrop fell—then another, and in a few minutes we were in the midst of a regular tropical downpour. Deafening thunder pealed almost incessantly, and when it paused nothing was heard but the loud swish of the falling rain. In a few minutes my clothes were soaked and the fire drowned out. We sat huddled together—I had my rug over my head to help to keep out some of the myriads of mosquitoes. The Indians too had doffed their solitary garments and put them over their heads. Hour after hour went by and it rained and thundered as much as ever. After a little I realised that I was sitting in a pool of water, but it would have meant too much to attempt to move. The water deepened, until about midnight the rain suddenly stopped, the moon appeared and then I understood the Indians' objections to my camping spot, for we were sitting apparently in the midst of a wide lake—all the ground for a long way round being submerged to a depth of about ten inches. The duck and the bits of deer projected in a mournful circle from the water close at hand. A fresh south-west wind now blew and the cold was terrible. The Indians, stoics that they were, wished just to sit it out till morning but I insisted on having a fire lighted by hook or by crook. Happily a little way off we found a small spot where the ground was not covered by the water and thither we moved our things. The problem now was how to light a fire and the Tobas were equal to the occasion. Sautranukdyî disappeared into the monte and we heard the twigs crackling as he moved about. After about 20 minutes he reappeared bearing a spherical termites' nest about a foot in diameter. This, constructed of gnawed wood and quite waterproof, afforded when broken open a supply of fuel, not very good but at least dry. With the help of the firesticks a little bit was set alight. Yorichidyî was then set to blow, which he did with all the strength of his lungs, and after about half an hour the fuel blazed up. Wood was then brought, first fine twigs, then thicker bits, and an immense pile was made which as it dried burst into flame. Lyaniteroi recovered the duck and about 1.30 we had an enjoyable supper. We sat round the fire warming our chilled bodies until too sleepy to remain longer awake and then we lay down round the fire.

Our legs and the lower parts of our bodies were lying in two or three inches of water, but our heads and shoulders rested on ground that was merely wet.

24th January: This morning it was with difficulty that I moved at all, so utterly stiff was I with sleeping in the water. We started off at once, and as the sun rose in the sky it became very hot and soon dried my clothes and drove the stiffness away.

A long day's march from dawn to nightfall, our direction on the whole easterly. The facies of the country is noticeably different from that round Fortin Page. The soil is richer and less impregnated with salt. For the most part the country is parklike, fine grass lands with many patches of monte. The montes too are richer than those further inland. The small monte fan-palm (*Trithrinax brasiliensis*) is frequent and now in full flower. The tall feathery leaved Pindó palm (*Cocos australis*) has also begun to appear in the montes, also the Bottle tree—Palo borracho (*Chorisia insignis*). In some places the grassy plain between the montes is dotted with numerous Ñandubeys (*Prosopis nandubey*), small trees with rough bark, and bipinnate leaves. Being now in the country of the Mushcui we avoid attracting attention by firing shots and feed mainly on honey and locust beans (Algarobo—*Prosopis juliflora*). The wasp honey Katik or Lecheguana is abundant and we eat it in large quantities. We are not troubled about water as pools are scattered everywhere owing to the recent rains.

25th January: Made good progress and in the early afternoon we crossed the Rio Negro, a small stream waist-deep of fresh muddy water. We had seen a column of smoke some miles distant to the westward and the Indians interpreting this as a Mushcui signal to gather in pursuit of the strangers within their territory, we set fire to the thick brush on each side of the river so as to obliterate our trail and hold up the pursuing Mushcui. Turning off downstream for a couple of miles, we then resumed our easterly direction and passed through a belt of dense forest. On its far side we encountered the worst bit of country so far, and all afternoon was spent in traversing miles of marshy flats, with tufts of stiff coarse grass separated by pools of water and mud into which the mules sank at every step. In a very short time I had to dismount and struggle on as best I could on foot, laden with the saddlebags, and dragging the reluctant mule behind me. Towards dusk we fortunately reached a slightly drier spot where we stopped for the night. Just as we were composing ourselves to rest, a rattlesnake sounded close by and Lyaniteroi chucked a small stick into the air and pronounced a short exhortation to the evil spirit to be gone. Though soaked to the waist in mud, I soon fell asleep.

26th January: Roused very early by the attacks of myriads of Polvorinos (*Ceratopogon* or *Culicoides*). Progress was again slow and difficult, through the same kind of marshy ground as yesterday. While traversing the country of the Mushcui we had so far avoided attracting their attention by gunshot, but this morning I shot a cierbo. We also saw fresh trail of a herd of Kuss, the White-lipped Peccary, but did not get into touch with them.

About 2 p.m. a real excitement—a carreta or waggon trail. Amongst the Indians I had of course been living in the pre-wheel age—is not the wheel one of the very greatest of man's inventions?—and it was strange to see again the ground marked by the two parallel furrows betokening the passage of something on wheels, and incidentally telling us that we were now on the verge of civilisation.

27th January: We delayed a little in the morning as Chimaki went off to hunt mañik, of which Lyaniteroi had seen several last night. He used the regular Indian technique [described on page 123], making a conical extinguisher-like construction of thick foliage and twigs which he settled on his head like an enormous hat, so as to conceal his head and the upper part of his body. He had, however, no success, so we continued on our way, following the wheel track except in places where it traversed low-lying country now under water owing to the heavy rains. As the morning wore on, signs of civilised man became more and more evident. We came in sight of herds of fine cattle grazing in rich pastures varied by islands of monte. The latter too were very different from the scrubby montes of the interior, being composed mainly of large trees, Quebracho predominating, with occasional Guayacán and Palo borracho.

About mid-day we stopped for breakfast and then pushed on. About 2 p.m. Lyaniteroi, who as usual was leading the way, signed to me to pass on ahead as we were close to the abode of the Cristianos, and sure enough turning a corner of the monte we found ourselves close to the Estancia house of Don Pedro Gil. A number of men were working outside but on seeing the unpleasant sight of armed Indians fled precipitately indoors and presently the barrels of rifles appeared—so I halted the Indians, dismounted and laid my rifle on the ground, and advanced alone to the houses. I found everybody with rifle in hand, many of them pale and trembling violently—though there were at least

15 men visible. They explained that they had been just on the point of opening fire on us, for they had seen no wild Tobas since their raids of by-gone days. The Estancia people naturally did not recognise me in my ragged European clothes and thought I might be the fearsome Cacique inglés, of whom they had heard rumours. However, all was well. There was much questioning about our expedition, mate was served and after a short pause we passed onwards on our route towards the Paraguay, I leaving my rifle and ammunition behind with the manager of the estancia. We lost our way twice and were further delayed by having to cross the deep and rapid Rio negro, so night came before we reached our destination. We kept on for a bit through swampy difficult ground but at last had to come to a stop, the mules being dead beat. We lay down where we were, without making a fire, and had a few hours of uneasy rest.

28th January: In the morning we reached the estancia of Don Diego Fernandez on the West bank of the Paraguay River, about half a mile above Asunción. Here we made a great sensation, the men advancing to meet us armed with every kind of weapon from 16-bore shot-guns upwards. Nerves were obviously excited so I decided to go back for some little distance and wait there until a canoe was available for crossing the river to Asunción. I also decided that it was not feasible to take the whole party of Indians to the city, and that all but Chimaki had better retire some distance into the Chaco and await my return. Lyaniteroi did not see the force of this at all, and it was only after I expatiated on the prevalence of smallpox in Asunción that he became agreeable to Chimaki alone accompanying me.

Chimaki and I then passed back to the estancia, unsaddled our mules and turned them loose, and settled down to await the arrival of the canoe from Asunción. A delicious cup of coffee and a pleasant chat with an intelligent Frenchman who had been an intimate friend of the explorer Thouar, and then at last we embarked and were paddled across to Asunción.

We were naturally much stared at, and the British Consul—Dr Stewart—equally naturally—failed to recognise a disreputable looking creature burnt nearly black with the sun, with untidy beard and moustache and clad in ancient and ragged garments covered with mud, as the naturalist of the Page expedition [Plate XVII]. However, matters were soon explained and the Consul was most helpful, his brother, Mr George Stewart, offering to put me up and lending me civilised garments until I should be able to purchase new ones. Money threatened to be a difficulty.

A clerk sent to the Banco del Paraguay reported that I had been presumed dead, my account closed, and my papers sent down to Buenos Aires. However the Consul advised me what to do—to draw a Bill on Buenos Aires or London for the money I wanted, which he would endorse and it would be at once cashed. And so it happened, and it was an interesting experience to find that the Bill on London realised more than its par value—owing of course to the fact that money due for payments there can be transmitted in the form of paper at the mere cost of postage—quite negligible as compared with that of transmitting gold at bullion rates.

Our five-day stay in Asunción was busy and interesting. Mr Stewart had a trestle bed rigged up for me in his verandah and Chimaki slept close by on a rug on the floor. He seems to enjoy living in a house for the present but he will on no account let me out of his sight for a moment. He follows me closely everywhere and when Mr Stewart proposed to take me for a walk and leave him behind he would have none of it. So we took him with us to a café and introduced him to ices and other delicacies—which he liked, with the one exception of coffee which he found disgusting.

In the streets Chimaki attracted much attention and the women in the market were particularly demonstrative, one seeming to wish to carry him off bodily. He also excited great interest in the Kosmos Club where he, like myself, was entered as an Honorary Member.

On the day after our arrival we returned to the west side of the river, where there were rumours of trouble. Indians were reported to have been prowling about the settlement all through the night and arrows had been heard whizzing overhead in all directions. Eventually they accepted my assurances that arrows don't whiz and that the rumours put about were undoubtedly mere fairy-tales.

One day I devoted to a journey down river to the Prefectura of Puerto Pilcomayo—the Argentine frontier post at the mouth of the river. There I was entertained to a sumptuous breakfast and discussed matters with the Prefect, Sub-prefect and Candioti, who were all very friendly and polite. Finally it was arranged that I should have a cavalry escort for my return journey as well as mule transport for supplies of provisions, which I proposed to purchase for our party on the *Bolivia*. It was made very clear that all this was being done to oblige myself personally and very unfriendly sentiments were expressed about Nelson Page.

Back in Asunción I arranged the purchase from Don Pedro Gil of a supply of bullocks, and from Messrs Uribe and Co. of other stores for my companions up-river.

On my last day in Asunción I was received by the President of the Republic who had already sent kind greetings. He made many compliments, asked many questions about the Indians, and referred to the difficulties with the Argentine over the boundary line separating the two Republics in the Chaco region.

I was much impressed by the President's remarkable interest in my doings, which interest however became more understandable when the Consul told me that the father of His Excellency, Don Juan G. Gonzalez, was reputed to have been a Scot of the same surname as myself!

Another thing of interest during my stay in Asunción was an invitation through Dr Stewart to take over a tract of country between the rivers Jejui and Aguara-y and colonise it with my Tobas. As a beginning I would have 30 leagues of well-watered country rich in valuable timber. Equipment would be provided—carts, bullocks, agricultural implements and 20,000 dollars cash. Development would be agriculture—maize, mandioca, mani, etc. —and timber. And I should be allowed to stick to half the profits. A tempting invitation which would have to be carefully considered!

Our last afternoon was devoted to shopping—axes, knives and so on as gifts for the Indians—shirts and other garments for myself. Mr Stewart bought a fine knife for Chimaki which gave him great joy.

Finally on the morning of 3rd February I completed my purchase of stores—rice, flour, coffee, salt, matches, etc., and we embarked on a boat which I had hired for the passage across the Paraguay. On the western bank we were delayed a good deal by having to catch and saddle the mules which had been running wild during our absence. When at last driven into the corral and lassoed they resisted violently all attempts to get the bit into their mouths. However the men were 'muy vaqueano' and had the mules down on the ground in a trice and the bits were soon adjusted.

I wondered what Don Diego would charge us for all the help

he had given with the mules, arranging transport across the river, and so on—but he firmly refused all remuneration and merely gave us his blessing and wished us a good journey.

Early in the afternoon we reached the Estancia Gil, having had a little trouble in crossing the Rio Negro which was in high flood. Corporal Gomez arrived from Puerto Pilcomayo with his party of soldiers and I arranged to take over the bullocks on the following day. I was hospitably entertained—though rather embarrassed by the persistent questionings by the local ladies regarding the matrimonial affairs of the Tobas.

It was nearly sunset next day before I could make a start. First there was delay in rounding up the bullocks. Then the Capataz or Mayor domo insisted that as our expedition was an Argentine one we must cross to the other side of the Pilcomayo instead of taking the much shorter route through Paraguayan territory. I on the other hand insisted on his agreeing to our taking the latter route. As he was quite unaffected by my arguments I finally told him I would proceed as I thought best. He was wise enough not to use force—which might have started endless trouble—so I wrote and signed a letter to him saying that I was taking the liberty of passing through his demesne and promising to keep to the track while so doing.

We stopped for the night only a league or two from the estancia house and next day made good progress through fine parklike country—again crossing the Rio Negro which now had hardly any water in it. In the evening we were joined by my party of Tobas and there was a vast amount of talking as Chimaki regaled his companions with an account of his experiences in the city of the Cristianos.

Most of the next day we spent coasting along the edge of a large estero. On this journey I fit in with the ways of my soldier companions. As dusk approaches and we arrive at a suitable camping spot we stop for the night, unsaddle our mules and get down our saddle bags. The mules are turned loose, one or two being tethered or hobbled. A fire is kindled, water is fetched and cooking commences. One of the men prepares the mate, which we take 'á la bombilla' i.e. sucking it up through a silver tube from the gourd or mate which is handed to each in turn. The infusion is very strong, of a deep green colour. Although an acquired taste, this Paraguayan tea—made from a species of holly, the Yerba mate (*Ilex paraguayensis*)—is a very valuable drink, acting as a gentle stimulant to the nervous system. Its flavour is not unlike that of the coca leaf from which cocaine is obtained. Morning mate, like the chewing of coca leaves, is an admirable preparation for a prolonged day's exertion, but unlike coca it leaves no unpleasant after effects.

While seated round the fire enjoying our mate one of the men prepares a stew of fresh venison with rice and pumpkin: another prepares soup of similar materials and we soon enjoy our repast. After dinner I 'turn in' lying on the ground with rug and coat and my saddle as pillow—nothing above but the clear sky with the Southern Cross and its myriad other jewels. Around one hears the occasional far-off cry of some nocturnal mammal, the croaking of frogs, the bark-like voice of the small Chaco tortoise, while close at hand the Indians converse in low and characteristic accents. Round the bright fire sit the soldiers and one of them has brought his guitar and gives us a concert of the low and sad-sounding impromptu melodies so much affected by the Spanish race. Which reminds me, by the way, that I have not mentioned my delight in Asunción when an excellent military band played over airs of my own country—our National Anthem, a highland march, the Last Rose of Summer, Home Sweet Home.

Two more days and we arrived (8th February) at the *Bolivia*—to find that once we had got away on our journey to communicate with headquarters the soldiers had been, as we expected, on their best behaviour.

A new Commander arrived on the 24th February—a Captain Bouchard. He had been in command of the first relief expedition which approaching the Pilcomayo from the south had proceeded naturally along the southern bank and being unaware of the existence of the northern branch, upon which was the *Bolivia*, had failed to find any trace of us. Incidentally he cleared up the mystery of the distant firing heard from the *Bolivia* on 1st June—on which day he had a fight with the Indians, entirely unaware of our being only a few miles off.

Bouchard was a seasoned Chaco soldier and gave me graphic accounts of his experiences. He had had, he said, thirty different commissions in the Chaco and not one with less than a hundred Indians killed. His stories of cutting wounded men's throats and ripping up women who tried to protect their nearest and dearest were extraordinarily cold-blooded, and harmonised unpleasantly with what the Tobas had told me about what was apt to happen when one of these expeditions of the Cristianos invaded their territory.

During my closing weeks with the *Bolivia* the aftermath of our raid upon the Orejudos made its appearance in the form of a series of reconnaissances and threats of attack. One night about nine o'clock the sharp ears of our Indians detected the sounds of stealthy movement on the other bank of the river. The women hastily gathered up their household goods and with the children popped into the neighbouring monte. We stood to arms and a party crossed the river to reconnoitre. They found nothing, but after a bit there suddenly came to us the alarm cry of a pair of chajás in the distance, telling us that indeed there were strangers about. And next morning I went off with Chinkalrdyé and Netalurdye to look for trails in the direction from which the sounds had come, and sure enough within a few minutes we found in the soft ground by the margin of the estero the fresh tracks of two Indians which Chinkalrdyé at once pronounced to be made by Mushcui scouts. Later in the day one of our Indians who had gone off a couple of days earlier to the north-eastward to hunt returned in a great hurry to report that the Mushcui were only a short distance off, that they had pitched their tolderia and were keeping us under constant observation on the look out for a good opportunity to attack, and especially to carry off our livestock which were at present 3 or 4 leagues away from the ship.

In the opposite direction there was also trouble, for one of Yordaik's men came in with the report that he and three others had been surprised by a party of Bermejo Tobas one evening when playing dice by the edge of the monte and he alone had escaped to tell the tale.

In the meantime life at the *Bolivia* went on much as before. In the intervals between collecting and hunting with the Indians I occupied myself in preparing for my departure as I was summoned home by the failing health of my aged father. Much time was taken up in soldering together tin boxes in which to pack specimens preserved in spirit as well as my entomological collections. The main collections were to remain on the ship and arrangements were made for their transport to Europe as soon as the *Bolivia* got down-river.

There was still no sign of the onset of a real rainy season: there was only a bare trickle of water in the river and the *Bolivia* progressed merely by slow steps from dam to dam.

On 28th February we reached Dam II: burst through it on 5th March but in less than two hours we were again high and dry. An additional dam (I*a*) was now constructed a little lower down. That we burst through at 7 a.m. on 12th March but by 10 a.m. we were again at a full stop—and there, on 15th March, I bade my final farewell to the *Bolivia* (Plate XVIII).

I had with me a party of four—Sergeant Diaz, Corporal Zapala, Cadet Lopez, and Villegas—a set of real toughs who would obviously deserve watching during our journey.

It was a touching farewell to my Natokoi. Nomatena, Maidukna and Coche wanted much to come too and so also my men comrades whose lamentations at being left with the Cristianos and expressions of hate towards them made me feel that the future was rather doubtful.

The Paraguayan government had sent me a permit to take my party through their territory which meant considerable shortening of the distance we had to travel. We did it in four days and the journey to Puerto Pilcomayo was quite uneventful.

Return to the outer world and civilisation was not unmixed joy. I had become inured to hardship and the absence of all luxury. I had become accustomed to live simply and dangerously and I had learned to love these Red men in whose hands my life had been during all these months. I did not suppose I should ever see them again and I could not but fear that their splendid race was doomed to vanish from the scene at no distant date as had already been the case with countless other tribes of Red men in the two continents of America.

There now followed a delightful holiday of three weeks in Paraguay, most of it spent at Villa Rica, a charming little town at what was then the terminus of the Paraguay Central Railway. A party of Scottish engineers were engaged in prolonging the line towards the upper Paraná and they had sound ideas as to the treatment to be prescribed to a young man who had been subjected to Chaco conditions for over a year. The charming hospitality of Mr Angus and his colleagues at Villa Rica left abiding memories.

Round the town of Villa Rica were many beautiful rides. The roads wound about through woods and open country, across plain and between low hills, the deep red of the soil showing in strong contrast with the rich green of the vegetation which walled in the pathway on each side. The roads were really sequestered lanes and riding along in the early morning one met numerous parties of Paraguayan women on their way to market, bearing on their heads their wares—a basket of fruit, a bundle of mandioca, or a jar of milk—each one clad in the usual Paraguayan garb, a tipoi or loose chemise of snowy whiteness cut low and embroidered in black round the neck, a white shawl over the head, a big cigar in one corner of the mouth, feet and legs bare.

Good looking—though not to my eye rivalling some of the Tobas: but perhaps this might be partly due to the beautiful figures of the latter being better displayed to view than those of the Paraguayas.

However, these women with their tipois, their strongly marked features, their flashing dark eyes, their magnificent black silky hair reaching in many cases to well below the waist, their demure 'adios' as they passed, made a characteristic Paraguayan picture.

Every here and there by the roadside a rude wooden cross, sometimes within a little enclosure, and sometimes entwined in a band of lace or muslin, kept alive the memory of one killed in the Great War in which little Paraguay defied her big neighbours Argentina, Brazil and Uruguay for no less than five years.

The country as compared with Chaco or Pampa seemed thickly populated, little huts dotting the countryside, each with its little enclosure, its mandioca, its pumpkins, melons and bananas—not to speak of its invariable zone of orange trees, planted it was said in the time of President Lopez who made it compulsory on all citizens to have a certain number of orange trees round their dwelling—no doubt with an eye to food supply during the forthcoming war.

Returning to Asunción I finally, on 15th April, loaded up my gear at Puerto Pilcomayo on to the passenger steamer and was seen off with due ceremony by the sub-prefect. The voyage was uneventful, though my fellow passengers were rather tiresome with their bombardment of questions about the Pilcomayo expedition. At San Antonio we paused to load oranges and saw one of the most characteristic sights of Paraguay. We hauled alongside a rough mole constructed of palm trunks; at the landward end of which lay enormous heaps of oranges under a canvas awning. Along this trooped in single file about a hundred Paraguayan girls in their white tipois, each with the usual big cigar in one corner of her mouth, each carrying on her head a shallow basket of oranges. Demure they looked as they trooped along the stage with their eyes usually discreetly cast down but occasionally flashing a momentary glance. Amongst the general sallow complexions stood out one exceptional blonde with the most lovely complexion and features. As always their figures were well formed, their limbs firm and lithe correlated no doubt with their toilsome life: the men loafed about and listlessly watched their womenfolk work.

And so to Buenos Aires and home.

Of the collections made on the Pilcomayo expedition I had been able to bring away only as much as could be carried on my three pack mules so I had to select carefully, confining myself to specimens representative of the neighbourhood of Fortin Page—in particular birds, flowering plants, Indian weapons and equipment, and photographic negatives. Of these the birds were reported on in *The Ibis*, 1892; and the actual specimens presented to the British Museum (Nat. History); the plants in *Transactions of the Botanical Society Edinburgh*, 1893, and duplicate specimens presented to the herbaria at Kew and Edinburgh. The Indian implements are in the Museum of Archaeology of the University of Cambridge.

As regards the bulk of the collections, H.B.M. Minister in Buenos Aires reported that the *Bolivia* had eventually—in June—successfully reached the river Paraguay but that the fate of my collections was a mystery, which remained unsolved.

Part II

LEPIDOSIREN EXPEDITION

1896–97

CHAPTER IX

LEPIDOSIREN EXPEDITION

The years following the Pilcomayo expedition were spent at Cambridge, where the School of Zoology occupied in that science a position comparable with that of the Cavendish Laboratory in Physics. The spirit of the Cambridge School at that time was mainly morphological, i.e. it was concerned with investigating the form and structure of animals, and with the fascinating problem of how their diversity had come about in the course of evolution. An important part of the teaching of the School was devoted to embryology; in which the student was able to view with his own eyes the process of evolution at work: seeing for himself how a highly developed creature such as a bird or a mammal passes through a stage when the main features of its organisation are those of a fish: that, to put it shortly, a bird or a mammal goes through a fishlike stage in the course of its individual evolution. It will be appreciated how the fascination of such a study gripped the imagination of the student.

However, while working through this part of the Cambridge training I came to appreciate what seemed to be a weakness of the science of embryology which might well involve the risk of arriving at fallacious conclusions. This weakness lay in the fact that the builders of embryological science had been restricted in their material. The known facts of embryology had for the most part been gleaned from the study of creatures that happened to be easily accessible rather than from those which had lagged behind in their evolutionary progress and whose structure was on that account relatively simple and primitive. Comparative anatomists had shown how the adult structure of such old-fashioned creatures threw light into many of the dark corners of evolutionary history; it seemed entirely probable that similar expectations might be justified in regard to the study of their embryology. Cambridge had indeed already done great work in this direction, above all Francis Balfour's studies of the embryology of the sharklike fishes and Adam Sedgwick's upon *Peripatus*, a wormlike relative of the centipedes and insects.

Among the old-fashioned types of vertebrate which still linger on at the present time one of the most interesting is the group of Lungfishes (*Dipnoi*) with three distinct representatives in widely separated parts of the world (Plate XIX (*a*)). The first of these was discovered by an Austrian traveller Natterer (1836), in the region of the Amazon, and was named by him *Lepidosiren paradoxa*. Not long afterwards *Protopterus* from the rivers of Tropical Africa was recognised to be closely akin to *Lepidosiren*, and then after a long interval a third lungfish, inhabiting the Mary and Burnet rivers of Queensland. A remarkable discovery was now made, namely, that the curious tooth plates in the mouth of this Queensland lungfish were identical with fossils from Aust Cliff near Bristol to which had been given the name *Ceratodus*. The lungfish of Queensland turned out to be in fact a surviving *Ceratodus* which had persisted apparently unchanged right down the ages from Jurassic times to the present day.

Bones of many similar creatures have been found in ancient rocks in widely scattered localities and it is clear that the three types of lungfish existing in the world to-day are the survivors of an immensely ancient group of animals which at one time existed over the larger part of the world's surface. It is not merely the occurrence of fossil bones in ancient rocks that show the lungfish to be creatures of great antiquity, for the study of their anatomy shows them to be particularly old-fashioned in their structure, belonging to a stage of evolution far older than that of the highly evolved modern fishes. This is brought out particularly clearly by the construction of their bodies in relation to movement. The modern fish is adapted to shooting through the water with great rapidity: its body is beautifully streamlined, it exudes all over its surface slippery mucus to diminish friction against the water; it is provided with a large powerful tail to propel it forwards. It possesses an automatically regulated float in the form of an air-bladder which keeps it at exactly the specific gravity of the water in which it is swimming, so that it does not have to waste energy to keep it from sinking down into the depths or floating up to the surface.

Quite different the lungfish, which wriggles along; waves of flexure passed back along its body pressing against the water, and so driving the creature forward. It is able to transmit these curves back along the body owing to the muscle fibres which run along each side of the body being divided up into successive blocks which are able to contract in series one after the other from the head end backwards.

That this is the method by which the ancestors of all vertebrates moved is shown by the fact that even the highest of their modern representatives, such as a man or a bird, with its innumerable muscles running in varied directions in order to bring about the complicated movements of limbs and trunk, possesses in the early embryo only the serial blocks of longitudinal muscles along each side of the body as required for wriggling or rippling movement. The more highly evolved modern fish, such as the herring or cod or trout, still possesses when adult these blocks of muscle —familiar as the flakes into which the flesh of a cooked fish falls apart, but in these modern fish a further development has come about, in as much as the hinder tip of the body has become broadened out into the flat tail which serves as the main propellent organ. The lungfish, however, has not yet developed this flattened tail: the tip of the body still ends in a simple point.

Of the three types of surviving lungfish even a superficial inspection brings out the fact that the South American and the African lungfish are much more nearly related to one another than either is to that of Queensland.

As regards scientific knowledge it was *Ceratodus* that up to the date of my expedition was by far the most completely known. Not only as regards its adult anatomy; every stage of its development, after their first discovery by a Cambridge zoologist, W. H. Caldwell, had been worked out in minute detail by the German Semon and a number of colleagues. *Protopterus* was fairly well known so far as its adult anatomy was concerned, and it had often been brought to Europe, thanks to its habit of burying itself in the mud during the dry season and its ability to survive the journey in its unopened ball of mud. I well remember at a meeting of the British Association in Liverpool seeing one of these mudballs placed in water and the lungfish brought forth and quite inadvertently overhearing a distinguished Belgian savant remark to his lady companion 'Ce n'est pas la zoologie, c'est l'obstétrique'. Curiously however, nothing was known about the reproduction or embryology of *Protopterus*. *Lepidosiren*, though the first of the surviving lungfish to be discovered, was still the least known of the three. Practically nothing was known about its habits and absolutely nothing about its reproduction or the early stages of its life history.

In addition to the sharks and lungfishes there still survives yet another very ancient type of fish, the ganoid *Polypterus* of tropical Africa, and in this case too the early stages of the life history were completely unknown.

Such were the facts that inspired me to dedicate my research career to the endeavour to work out the embryology and life history of these three ancient types of Vertebrates—*Lepidosiren, Protopterus* and *Polypterus*.

Now during my undergraduate period at Cambridge there suddenly came the startling announcement that a German collector (Bohls), employed by the Hamburg Museum, had obtained several specimens of Lepidosiren from Indians on the River Paraguay, and on hearing this there suddenly flashed across my memory the tale of an eel-like fish which one of our soldiers reported having found buried in the mud of an estero on the Pilcomayo and my wondering at the time whether by chance this could have been a lungfish with habits like those of *Protopterus*. Together, Bohl's discovery, my memory of the soldier's tale, and the fact that I had unique experience of life in the Chaco, decided me to give priority to the Lepidosiren part of my programme and to proceed the moment I had got quit of my Tripos to organise an expedition to the Gran Chaco for the purpose of tackling the Lepidosiren problems.

While engaged in planning my expedition there came to me one day a friend who said 'John Budgett of Trinity is just dying to go with you'. I pooh-poohed the idea at first, for Budgett had no experience of roughing it in wild country, but his friends were persistent so at last I invited him to dine with me at the Café Royal and talk matters over. That talk, helped by an admirably cooked roast wild duck washed down by Clicquot—vin rosé, resulted in the settling of all my doubts, and the recruitment of a courageous, tough, and loyal comrade.

My plan was to make my way overland from Asunción to the country of my old friends the Natokoi and settle down among them and search the swamps for the lungfish.

The prospect of an arduous journey on mule back through the

Chaco entailed elaborate planning to ensure my scientific equipment being kept down to the minimum weight consistent with having available the apparatus and the complicated chemical reagents required for the most perfect preservation of delicate specimens destined to be investigated by refined laboratory methods. In order to prepare complicated solutions with the minimum of trouble I had their various constituents sealed up in thin glass tubes, in carefully weighed amounts. Then the preparation of the required solutions would merely require breaking the proper set of such tubes in the appropriate volume of water. Photographic materials were a difficulty. These were the pre-film days of glass plates; but the much lighter plates of celluloid had just been invented and I made careful inquiries as to whether it would be safe to depend upon them in the trying climate of the Chaco. Unfortunately I was advised by the experts that I should be quite safe in doing so and our entire outfit of photographic plates were of celluloid. The result was disastrous. Hardly a single negative was worth printing from and I was somewhat annoyed on my return home by friends marvelling at my ignorance of the need to guard against 'metal fogging'—a discovery made actually during my absence, that emanation from the tin in which the plates were hermetically sealed was capable of passing through the celluloid and thus ruining the sensitive emulsion.

An uneventful transatlantic voyage terminated in our arrival at Ensenada the port of La Plata on 29th August 1896. Buenos Aires had, I found, undergone a marvellous change since I had first landed in 1889—the swampy shore along which I had hunted birds being now replaced by a magnificent range of docks with all the latest equipment. A week in Buenos Aires was spoilt by a flare up of malaria which had remained dormant since the time of the Pilcomayo expedition. At least that was the medical diagnosis and it seemed probably correct in spite of the length of time involved, as the only alternative was a fresh infection at Rio and this seemed improbable.

On 6th September Budgett and I embarked on the *Olympo*, one of the fleet of fine river steamers created, and formerly owned, by a British company—La Platense—though now taken over by a Slav—Mihanovich. On board the *Olympo* was a Scottish fellow-passenger—one Graham—who told us he was bound for Paraguay as a recruit to the South American Mission working in the

Gran Chaco under the leadership of W. B. Grubb. At the mention of this name I pricked up my ears and remembered that Bohls in his note on the discovery of Lepidosiren in the Chaco had referred to the assistance got from 'mein Freund Grubb', and, as it seemed probable that this must be the same individual, I decided that instead of going back to the country of the Natokoi as originally planned it would be wise in the first place at all events to make contact with the Chaco missionaries though this would involve proceeding beyond Asunción a couple of hundred miles farther up the Paraguay to Villa Concepción where the Chaco Mission had its base.

At 6 p.m. on 13th September we tied up to the mole at Asunción, and blithely seizing our hand luggage we prepared to step down the gangway—only to be stopped by a sentry with a bayonet. This somewhat chilly welcome to Asunción was due it transpired to the need of protecting the ancient rights of the ferrymen who until the construction of the mole held a monopoly of landing passengers from the river steamers. The rule now was that a new arrival must embark in a canoe on the outer side of the steamer and make his first journey ashore by it: thereafter he was free to use the gangway as much as he liked!

This little difficulty surmounted, Budgett and I boarded a tram and proceeded to La Cancha—a hotel situated a mile or two south of the city on high ground with a wide view across the river into the Chaco. As we sat in the open air in the glare of an electric arc and ate an excellent dinner we could see on the far north-west horizon the glare of Indian fires—it might be of the dear old Tobas. Budgett was greatly thrilled as we sat eating our dinner and thereafter smoking our cigars under conditions of full civilisation, to realise that just across the comparatively narrow river was the Chaco 'infiérno' or hell as the Paraguayans called it still practically untouched by civilisation.

Ten days at Asunción were well occupied in shopping, completing our equipment—saddlery, top-boots, bombachos, mosquito nets, etc.—interviews with government officials, meetings with old friends and so on. Among these old friends were the charming wife and daughters of H.B.M. Consul-General—Dr Stewart.

A day's journey in the river steamer *Aurora* brought us to Villa Concepción, a small town which served as one of the chief ports for the export of the Yerba mate, or Paraguayan tea, brought down in convoys of bullock waggons from the distant Yerbales of the interior.

Nearly a month was spent in Concepción completing our arrangements and incidentally doing a good deal of collecting work. In the lagunas opening off the river we worked my 80-foot seine net—with difficulty on account of the number of submerged logs and portions of trees. Our hauls gave us a fairly representative collection of the local fish fauna—a great variety of the voracious Characinoids including the Palometa (*Serrasalmo spilopleura*), Catfish, a true Sole (*Solea jenynsi*), and an interesting Sting-ray (*Taeniura dumerilii*).

The Concepción station of the South American Mission was situated on the Chaco side of the River Paraguay, on the bank of a tributary—the Riacho Negro. On proceeding there we found that Mr Grubb was away on leave in England but we were able to discuss plans with his deputy Mr John Hay. From him we learned the all-important fact that the main work of the Mission was at the time being carried on at a station far in the interior of the Chaco called Waikthlatingmayalwa, among a set of Lengua (i.e. Mushcui) Indians known as the Paisiapto or black-food people, from the fact that their main food was a dark-coloured eel-like fish that abounded in the swamps by which they lived. It seemed quite probable that this creature was in fact the Lepidosiren or Loalach and that accordingly the thing to do was to scrap the Pilcomayo plan and proceed to the country of the Paisiapto. Mr Hay was good enough to make business arrangements whereby we should be provided with transport by bullock carts to carry our impedimenta on the journey, and thereafter with board and lodgings for ourselves. Thus through the aid given by these excellent missionaries the entire problem of the Lepidosiren expedition was enormously simplified from what had been my original scheme.

At the level of Concepción the River Paraguay is split into a wider eastern and a narrower western channel by a long island and our journeys to the Mission Station were either entirely by canoe round the north end of the island or more often by direct route, crossing each branch of the river by canoe and the island on foot. On one of these journeys the Lepidosiren expedition nearly came to an abrupt end.

We had just purchased a couple of horses and one of them gave great trouble as we swam him across the wide eastern branch, being apparently determined upon joining us in the canoe. In the western channel he was still more difficult and sundown was approaching before we at last got him ashore on the Chaco bank, and started on our journey back to Concepción. The sky was black and threatening and it thundered heavily away to the north-west so we did our best to get across the island before night fell. This was not to be, the storm closed in around us: thunder banged and lightning flashed continuously and rain fell heavily: and by the time we reached the eastern bank of the island it was quite dark. Seeing the fires of a Mushcui tolderia we went towards it and called out for a canoe (Plate XXII (*b*)). After a long delay an Indian, very queer in his manner, appeared and agreed to ferry us across. He had no paddle but we told him to fetch one, and in a quarter of an hour or so he reappeared with it, looking queerer than ever. He seemed muddled as to which canoe it was to be but at last signed to one into which we got. He seemed about to push the canoe off and we were very nearly swept down stream but at last he staggered into the canoe—a slender dug-out about 12 inches wide, into which we could just squeeze, sat himself down in the bow and we recognised that his queer behaviour was due to his being very drunk.

Slowly and erratically we proceeded out into the channel backwards in cuttlefish fashion, the broad flat stern throwing clouds of spray into the air. We were beginning to wonder if we should ever reach the far bank when looking upstream I saw by the lightning flashes a great white mass coming down upon us—a horrid wicked-looking squall. A terrific wind struck us and in a moment a sea was running which threatened to swamp us at once. Budgett was sitting in the stern and his broad back did much to shed aside the oncoming seas. Others toppled in over the sides and the water in the canoe rapidly deepened in spite of my hard work baling with a large tin basin which Budgett had kicked within my reach. It seemed utterly impossible that we could avoid being swamped, which with the strong current and heavy seas would necessarily have meant the end of the expedition. However, praise be, when the canoe finally submerged it was in shallow water close by the bank and we were able to wade ashore and proceed, soaked but otherwise none the worse, to our lodging and evening meal.

20th October: We loaded horses and baggage on a lighter and about 4.30 began to crawl slowly upstream, urged by a small steam tug lashed alongside, panting and making a great fuss. About 7 o'clock we had a somewhat frugal meal in the pleasant cool of the evening, and after a pause of a few hours, mosquitoes and midges being rather bad during the latter part of the night, we got under way about 7 a.m. and three hours later tied up at Carayá Vuelta the end of our river journey. Here we paused for a day at a small rancho built of palm stems and belonging to the mission.

Hunt, one of the missionaries, and Sibeth—their German stockman—were awaiting us at Carayá Vuelta with a couple of bullock-carts in which we packed our equipment and stores, and on the morning of the 23rd we saddled up and started on our journey westward, I riding the chestnut, Budgett the black, while our two spare horses ran free.

The countryside with its palmar, monte, and estero, showed little difference from the Chaco of the Pilcomayo region: the same birds, the same deer, and the same biting flies—horse-flies, mosquitoes and midges. The flies were at times very bad and the heat in the afternoon was trying: at one point my horse suddenly flopped down in a faint, overcome by the heat, so I had to wait for a bit to give him a rest.

Heavy rain fell at intervals and I began to worry about the imminence of the wet season when it might be expected that the Lungfish would begin their spawning operations. One of the Indians said that by now the loalachs would be making their 'houses' in which to lay their eggs. So I decided to push on to Waikthlatingmayalwa, ahead of the slow-moving bullock carts.

27th October: During the night we had an unpleasant experience. As darkness came on there were signs of an impending tempest. We made all preparations—having a hurried meal and packing all our spoilables into the shelter of the cart. Presently a dull roaring sound like the rushing of a mighty river and the storm was on us, the palms bending and swaying and the whole scene lighted up by the almost continuous lightning flashes. The wind having at last died down we thought we would make an experimental trial of a six-foot-square tent I had brought with me. We pitched it in the lee of one of the carts to which we tethered it with strong additional stays. The tent would just hold three of us so Budgett, myself and Sibeth composed ourselves to rest under its shelter. All went well until between 12 and 1 o'clock when I was aroused by Sibeth who lay next the entrance suddenly dashing out into the darkness shouting various unfamiliar expressions among which I could recognise 'las hormigas'—the ants. Realizing that our tent was being invaded by an army of foraging ants I jumped up and seizing bed and bedding chucked them out of the tent, not too soon however for simultaneously I felt innumerable fiery pricks all over my body, the ants having got into position under my clothing preparatory to their anticipated meal. Tearing off my clothes I devoted myself to detaching the unpleasant insects whose jaws were firmly embedded in my skin. Presently we were joined by Budgett when the invading hosts reached him, but he had not fared so badly as he was wearing his top boots which formed a barrier against the intruders.

About 4 p.m. Budgett and I left the others and started off for the country of the loalachs with our four horses, and an Indian 'Philip' to act as guide.

29th October: A frightful night, excessively hot, and myriads of biting flies. High fever during the night and no sleep. I imagined I had encountered a new and particularly virulent type of malaria but it turned out to be something quite different for when daylight came we found our camping spot alive with myriads of fleas, and the febrile symptoms were simply the direct results of their poison. No doubt we had happened to choose a camping spot previously used by a party of Indians who had left behind some of their fleas which had multiplied actively to produce a vast flea population equipped with peculiarly healthy appetites.

When at last day broke I felt a complete wreck. The two spare horses had vanished so we saddled up those we had ridden yesterday and, partly restored by a cup of Savory and Moore's peptonised cocoa, I managed to clamber up into the saddle. We kept on for a couple of leagues and then changed horses, Philip having succeeded in rounding up the two strays, while those we had been riding were completely done up by the great heat and the heavy going. About eleven we reached an old hut built by the missionaries and I at once lay down to get some sleep, rest being more needed than food.

About three we got under way for the last stage of our journey. The horses seemed to realise this for they went well in spite of the going being still very heavy with a foot or so of water over most of the ground.

A little after five the mission buildings of Waikthlatingmayalwa hove in sight, looking with their corrals and livestock—cattle, sheep and goats—like a large estancia.

Mr Andrew Pride—acting head of the mission—welcomed us and we proceeded to rig up our beds—of strips of hide in a rectangular wooden frame—under the verandah.

Close by were a party of Indians cooking their supper—of Lepidosiren!—and presently one of them brought along a plate of the cooked lungfish which I ate 'con mucho gusto', for the flesh, rich with its deep orange red fat, was most tasty. [Later on, after my return home, I was told how scandalized my old professor (Newton) at Cambridge had been on hearing of this sacrilegious use of the sacred Lepidosiren.]

My enquiries about the Loalach brought a most welcome reply—that they lived in great abundance in the neighbouring estero. Pride expressed regret that the heavy rains just begun had forced them out of their dry season burrows, but to me this was not regrettable but on the contrary most welcome, for the onset of the rains would surely mean the beginning of the spawning season and the most important object of my expedition was to investigate the development of the young and hitherto unknown stages in the life-history.

30th Oct.: In the morning after a look round the mission station—with its typical Chaco surroundings—palmar, patches of small tree monte, scattered Palo borracho, Vinal etc., with some fine Cacti and Prickly Pears—we proceeded with Pride to visit his garden on an island in the swamp, where he grew sweet potatoes and other vegetables. As we waded along, the water about knee-deep, there was a sudden cry of 'loalach' from one of the Indians who had discovered one of the wet season abodes of the lungfish already excavated in the bottom of the swamp. Unfortunately its owner was not at home but later on, as I prowled about in the swamp with Philip and a couple of Indian boys, a Lepidosiren was sighted and one arrow then another and another were let go at it and we were able to seize it in triumph transfixed by three arrows and snapping viciously. I shrieked to Budgett who was some distance off and after gazing at our beast for some time we set off homewards.

On reaching the station we put the lungfish, seemingly not much the worse for its wounds, into a basin of water and were able to see that it still kept up its ancestral habit of taking in water by its mouth and passing it through the gill-opening, but that it also breathed air directly, pushing the tip of its nose above the water at short intervals and 'swallowing' air into its lungs.

Soon after our return one of the Indians came in with five Loalachs—four females and one male. He had cut off the hind limbs of the latter. Why, oh why?

A Boa—about ten feet in length—was also brought in: its flesh quite good eating.

4th Nov.: A fascinating discovery to-day which probably gives a clue to the mysterious removal of the hind limbs from a male lungfish brought in the other day. Lankester had discovered that the hind limbs of male specimens collected by Bohls had curious little knob-like projections along their upper side, but the meaning of these was a complete mystery. Now in two males brought in to-day the hind limbs were greatly enlarged and the little knobs had grown out into long blood-red filaments [Plate XX (*a*)]. The Indians say the male remains in the burrow to protect the eggs and it seems clear that the modified limb with its long filaments, rich in blood supply, must act as a temporary breathing organ, so as to diminish, or obviate, the necessity of visiting the surface of the water to breathe air into the lungs, and so leaving the eggs or young open to attack.

7th Nov.: The carts arrived yesterday just before sunset so to-day has been busily occupied in unpacking and arranging our laboratory equipment. Our laboratory is to be under the veranda at one end of the missionaries' dwelling house. Here we have fixed up a rough table, with packing cases on which to sit and with bottles of reagents, microscopes, dissecting instruments and so on, all arranged so as to be readily available. Amongst our large amount of glassware and other fragile bits of equipment there is not a single breakage in spite of the rough jolting on the journey over-land, which emphasizes how well worth while is the expenditure of much time and care upon packing the equipment of a scientific expedition.

8th Nov.: I went off with a witch doctor on a fishing expedition. He took with him his loalach spear, a cane rod, line, and hooks and we waded away to the South-West to a spot where the water was a little over mid-thigh deep. The Indian bent the tall papyrus-like rushes apart so as to leave a clear pool. Then he took a large water-snail (*Ampullaria*), tore its softer parts into fragments which he scattered in the water to attract the fish while the tough muscular part he transfixed with his hook to serve as bait. The fish began to take almost at once, mostly the two common Characinids (*Macrodon trahira*, and *Erythrinus unitaeniatus*), but after he had caught 15 there came a pause. He still persevered and kept on for an hour and a half or so longer, I standing by all the while in water nearly up to the waist. As I waited patiently I could not help passing back in imagination to the early times of man's communal evolution and realizing that the witch doctor of those far off days was the common ancestor of our scientist, priest and physician. The witch doctor's authority rests primarily on his scientific knowledge of his environment—its natural phenomena and its fauna and flora—but with it goes the knowledge accredited to him of the supernatural and his consequent power over the evil beings that cause misfortune and disease.

At last my companion saw fit to make a move and we continued on our way, making a wide irregular curve through the further parts of the swamp. Once a sudden splash occurred close by and I saw a Lepidosiren hurrying off. Then later a slight quivering movement of the swamp grass was visible passing rapidly seven or eight yards away to the Indian's right hand. He followed it up but still the movement evaded him. Next time, however, he had successfully marked the exact spot at which the movement ceased and two or three sharp thrusts of his spear resulted in his transfixing a male lungfish nearly 80 cm. in length. With some difficulty owing to its extraordinarily slippery slime I managed to get it into my bag and we went on. Then we got a second, rather smaller, male lungfish and saw indications of seven or eight others without, however, any further success.

All these lungfish were in water mid-thigh to waist-deep with coarse grass very much like a flooded meadow. In no case did we see any in the peguaho, as the Paraguayans call the deeper parts of the swamp marked out by their tall Marantaceous plants and Heliconiums.

9th November: The witch-doctor was out fishing again and came in with a fine bag of a dozen loalachs. By the way when hunting loalachs the Indian has an excellent way of transporting his catch, threading the heavy lungfish on to a string passed through mouth and gill-opening and towing them behind him, the slime covering their surface causing them to slip through the swamp vegetation with practically no resistance. Only when he emerges from the swamp on to dry land does he pack the loalachs surrounded by grass into an open net which he carries slung over his shoulders.

One of to-day's loalachs, a large female, yielded a great discovery, for when the cavity of her body was opened there lay exposed to view—for the first time in embryological history—a considerable number of the beautiful pale orange-salmon-coloured mature eggs of *Lepidosiren.*

Each egg is spherical, one hemisphere more deeply coloured than the other, measures about 7 mm. in diameter and at this particular stage is enclosed in a clear, transparent, gelatinous-looking, colourless envelope, nearly a millimetre thick.

16th November: Last evening an Indian reported that he had found a loalach nest with eggs in it, so this morning I set out with him and a couple of Indian boys to inspect it. We found the nest with some difficulty but alas! the eggs within it were clearly not of the Lepidosiren but of the much less noteworthy swamp-eel, *Symbranchus*. We now proceeded to explore the swamp systematically, working towards the deeper water. Every now and then one of us would suddenly sink up nearly to the neck; he had slipped down into a lungfish nest. One of the Indians would then explore the cavity with his foot to make certain. Most of the nests were empty but suddenly a tell-tale quivering of the grass was seen as a loalach hurried away and presently we found not only the nest but that it contained eggs. At this point the water was a couple of feet deep with grass and the papyrus-like rushes growing up through it. The nest was an elongated burrow, eighteen inches by eight, its floor about nine inches below the level of the bottom of the swamp, provided with a wide opening at one end. It had no special lining, its walls being composed of the ordinary black mould full of grass stems and roots. The eggs, lying loose in the interstices of the nest floor, were only few in number compared with what I had found lying loose in the body cavity of an adult a week ago, so it looks as if the spawning were not completed at one operation.

Having collected the eggs from the nest I carried them to the station in one of the enamelled iron pitchers which I had provided for the purpose and the rest of the day I spent in the fascinating, but very delicate task of extracting the eggs from their protective envelope. This latter had now taken the form of a thin horny shell which had to be cut round with fine scissors, the egg being kept submerged the while in water. The actual embryo extended nearly completely round the egg, the main mass of which was composed of slightly coherent granules of yolk which gave the salmon colour to the egg. It was clear that to get a really complete view of the embryo it would have to be separated from the yolk and extended out into one plane. To achieve this successfully proved a most grisly piece of work as the substance of the embryo was so delicate as to disintegrate on the slightest provocation. Each embryo had to be floated with the greatest care on to a little square of thin glass, brought up nearly to the surface of the water in the glass dissecting dish and carefully flooded from a pipette with a preserving fluid which would coagulate the protoplasm and give the embryo a consistency sufficient to permit its being gently raised out of the water for transference to spirit without disintegrating. However, I felt amply rewarded for all the trouble I had taken in carrying out this operation by the sight of the beautiful embryos, each extended flat upon its square of glass and ready for the subsequent stages of laboratory technique.

Altogether it had been a great day's work but it was not yet complete, for one of the Indians brought in a second batch of eggs which proved to be at an earlier stage of development without as yet any trace of obvious embryo.

From now onward there ensued a period of continuous laboratory work, my normal day being spent in my hut laboratory, researching into the material brought in by the corps of Indians whom I had enlisted as collectors, and putting it through the various methods of treatment to ensure its arriving at the Cambridge laboratory in the most perfect condition for detailed investigation. It will therefore be convenient to forsake for the time being the diary form of record.

As already indicated, the Gran Chaco forms an almost dead level plain with a gradual slope to the south-east, this slope determining the direction of its drainage towards the Rio Paraguay. The slight variations in the height of its general surface are marked by conspicuous differences in its vegetation. The deeper parts of the swamp where the water is practically permanent are marked by tall Marantaceae and Heliconiums conspicuous from a distance by their broad green leaves. Next come the main portions of the swamp, completely submerged in the wet season to a maximum depth of it may be 4 or 5 feet in a normal season. Slightly higher in level is the palmar, still liable to occasional flooding as shown by the flood marks on the palm stems; and finally higher still the level of the monte where flooding does not normally take place.

The typical swamp which forms the home of the lungfish comes in the second of these categories. Its appearance is that of a great meadow, being for the most part overgrown by a dense mass of coarse grass, varied here and there by beds of tall papyrus-like rushes or by spaces of open water dotted by the leaves of water-lilies.

With the first onset of the rains, normally in the early spring but by no means regular, the loalachs break their way out of their dry-season burrow and take up their free life. Their favourite food is the great water-snail *Ampullaria*, but they also eat vegetable matter. The separate teeth of the young stages become joined up as development proceeds so as to form in the adult large ridged plates admirably adapted for masticating the food.

Breeding soon begins, the eggs being laid, as already recounted, in deep burrows at the bottom of the swamp. The male parent remains on guard and in this connexion came the interesting discovery already alluded to, that the hind limb becomes temporarily a great gill or breathing organ so as to obviate the necessity of deserting the eggs in order to repair to the surface to breathe air. When his work of guarding the eggs and young is finished and he resumes his free life in the swamp the gill filaments rapidly degenerate and disintegrate and the limb reverts to its normal condition with little rounded knobs representing the vanished filaments and ready for regrowth when the next breeding season comes round.

The early stages of development of the lungfish are of merely technical interest but eventually the egg develops into a somewhat tadpole-like larva with a greatly swollen front part of the body containing the yolk on which the creature will subsist up to the time that active feeding begins (Plate XX (*b*)). At last the larva breaks out of the egg shell and now came an interesting discovery, for the shell was found to undergo a preliminary softening, a kind of digestive process through the action of a ferment secreted by the skin. Since this discovery in *Lepidosiren* similar hatching ferments have been found to be present in various other types of animal.

The young larva is entirely devoid of the pigment which gives the adult its dark, almost black, colour. The first to appear is in the eyes which become conspicuous as two black spots. Then other scattered cells develop granules of dark pigment in their protoplasm and these cells show movements which betray their sensitiveness to light, creeping towards the upper side of the animal and settling there so as to form a light-proof roof over the living cells of the body.

The larva gradually loses its tadpole shape as the yolk in the front part of the body is used up, but before this happens there make their appearance on each side just behind the head four little knobs one behind the other (Plate XXI). Each of these develops into a beautiful feathery gill and when the red haemoglobin develops in the blood one can see with the help of a magnifying glass the stream of red blood corpuscles passing up and down the gill filaments with a jerky movement in response to the beats of the heart. These external gills, like the similar arrangements on the hind limb of the adult, are merely temporary, shrivelling up and disintegrating when the end of the larval period is reached.

Later on I found that external gills exactly similar in their mode of development are present also in the young stages of the African lungfish *Protopterus* as well as in the ancient ganoid fish of tropical Africa, *Polypterus*. It was their occurrence in these three ancient types of fish as well as in the most primitive amphibians—the newts and salamanders—that forced me to the conclusion that they are organs of great antiquity, inherited by the vertebrate animals of to-day from far back common ancestors.

One of the fascinating problems connected with the evolutionary history of vertebrates has to do with the appearance in them of a new set of organs to support the body and enable it to move along a solid surface. These are the two pairs of legs—fore and hind. Where did they come from? Out of what pre-existing organs did they evolve? I believe that the spread of knowledge regarding the development of the external gills of *Lepidosiren* and of these other relatively primitive vertebrates that I have mentioned will bring general acceptance of the view that they provide the solution of this problem.

So far the strange metamorphosis of the hind limb of the male Lepidosiren during the breeding season has been discussed in relation to its functional significance as a respiratory organ enabling its owner to extract the needful supply of oxygen directly from the water in spite of the fact that its internal gills—the normal breathing organs of ordinary fish—are entirely degenerate and useless, and in this way to be relieved from the necessity of deserting his charge while proceeding to the surface of the water to inhale air. The temporary changes in the appearance of the limb is, however, also of great interest to the student of evolution, for upon my theory of the evolutionary origin of the limbs it becomes explicable as a simple reversion to a far back ancestral stage, when the prospective limb was still a mere external gill. On this theory both pairs of limbs have evolved out of external gills and I was accordingly delighted to find that an occasional male Lepidosiren turned up in which the fore limb also showed a distinctly recognisable though not so pronounced reversion towards the external gill stage.

It would seem that the external gills are really the surviving remnants of a row of such organs originally present along each side of the body of the wormlike ancestral vertebrate. Most of them have vanished without leaving a trace, their breathing function having been taken over by internal gills, safely ensconced within the gill openings as they are in any ordinary fish, but of the original series two pairs have been saved from disappearance by their having taken on a new function—that of moving the body by pushing against the solid substratum—and it is these that have in the course of time evolved into the limbs of modern vertebrates.

At their first appearance the limbs of *Lepidosiren* are in fact little knobs identical in appearance with the first beginnings of the external gills. They gradually increase in length and become the simple fingerless and toeless limbs of the adult. The young Lepidosiren does make actual use of these primitive limbs and it was interesting to notice that the larger hind limbs when used for pushing the animal forwards took on a double curvature foreshadowing the knee and ankle joints in the hind leg of a land vertebrate.

In point of fact the two other lungfish show that the evolutionary story is not so simple and straightforward as might be inferred; for in the case of *Ceratodus* each limb is in the form of a clumsy, fleshy, pointed paddle supported by a characteristic skeleton while *Protopterus* shows slight but recognisable vestiges of a similar condition. Even in *Lepidosiren* the fore limb occasionally shows the remains of an expansion along one side. So there is no escaping from the conclusion that the limbs of the ancestral lungfish were paddles like those of *Ceratodus* and that the limb of *Lepidosiren* is not a persistence of but rather a reversion to the still earlier rod-like or styliform limb which was the first stage in the evolution of legs in those vertebrates which took to moving on a solid substratum. Nevertheless, the limb of *Lepidosiren* is of fascinating interest as giving us for the first time a picture of what the legs of land animals were like in their first beginnings.

As regards the typical fish in which the limbs have taken the form of thin membranous fins adapted to a purely swimming existence, it is again of great interest to find that in the most ancient types of these still surviving—the sharks and dogfish—the study of early stages of their development has disclosed a certain amount of evidence to show that there too the present-day fins have evolved out of an ancestral type like that of the present-day *Ceratodus*.

The facts discovered during the Lepidosiren Expedition followed up by later laboratory studies on the material then collected, as well as on the material of *Protopterus* and *Polypterus* collected by Budgett, and reinforced by what we know to-day of the development of other types of fish, lead, as I believe, to important conclusions as to the early stages of the evolutionary history of modern vertebrates.

The ancestors of this great group of animals were on this view creatures that inhabited the marginal zone of water immediately bordering upon the dry land. To the primitive mode of movement by eel-like wriggling they had added the new method of pushing themselves along by simple styliform legs evolved out of external gills.

The descendants of such ancestors gradually diverged along two different routes. The one of these led to a purely swimming existence away from the solid land, in adaptation to which the styliform limbs became flat fins. This line of evolution gave rise to the various groups of fish. The second route led to a purely land existence and in adaptation to this the styliform leg became modified into the typical legs as seen in such groups as reptiles and mammals. The group Amphibia, including the more primitive newts and salamanders and the more specialised frogs and toads, represent a less successful attempt to become land animals for they still demand a moist atmosphere, and in most cases actual water in which to pass through the early stages of their life history.

The larva of *Lepidosiren* with its fully developed external gills is a beautiful little creature. While at rest it lies among the grass and other debris its head end protruding and its bright red feathery gills spread out on each side, and every now and then it will flick its external gills sharply backwards, no doubt for the purpose of renewing the water in contact with the surface of the gill. Apart from its functional significance, this movement of the external gills disclosed two important secrets: (1) that the gills are equipped with muscles to produce the flicking movement, and (2) that the movement of the gill causes the whole larva to jerk slightly forwards. In other words, the external gills with their equipment of muscles is seen to possess therein, the necessary qualifications for evolving into a limb by increase in their power of movement and decrease in their function of breathing. Here we are brought into touch with a great evolutionary principle, that of 'substitution of functions' where a function that was originally subsidiary comes to be the predominant one while the formerly dominant one fades away.

An interesting thing about the Lepidosiren larva is the development of a peculiar cement gland which, arising as a crescentic structure just behind the position where the mouth will appear later, becomes a prominent cushion-like organ that secretes a sticky cement by which the larva hangs on to the material forming the floor of the nest and in this way saves itself from getting lost by sinking down among the loose debris. Like the external gills, this cement gland is merely temporary, being completely absorbed as the young animal becomes more active and able to look after itself.

Presently the young Lepidosirens are seen to come to the surface of the water and take in air into the now functional lungs and as these become able to cope with the entire respiratory needs the external gills assume a more and more unhealthy appearance, their circulation deteriorates and they gradually degenerate and finally disappear without leaving a trace.

About the same time, the end of the larval period, the young Lepidosirens show other changes. They become much more active, swimming like an eel in the primitive method by waves of flexure passed back along the body. On occasion they can reverse the direction of these waves of flexure so that they shoot rapidly backwards.

They show great sensitiveness to vibration in the water, a slight tap on the dish containing them converting its contents into a wriggling mass of activity.

The pigment cells have meanwhile greatly increased in number and crowded together close to the surface of the body so that the young fish is much darker in colour—nearly black—as compared with the larva. This crowding together in the skin is to shelter the living protoplasm of the body from the light, for it is an undoubted fact that rays of light are harmful to living protoplasm, a fact the appreciation of which tends to be obscured through so much of man's environment consisting of green plants dependent upon light for their nutrition, and again by the potent psychological effect of bright sunny conditions.

The young Lepidosirens also showed their extreme sensitiveness to light by their wild commotion when the wooden cover was removed from a vessel containing them, and their tendency to seek the shade whenever it was available. Their relations to light and dark came out in a more startling way in a discovery on 25th January. A disaster had happened through the fowls of the Mission Station invading a shelf on which I had my dishes containing the young Lepidosirens and capsizing the dishes. Two-thirds of the 150 or so were dead and the remainder I rescued and conveyed to a place of greater safety. About 9 p.m. I took a lamp and proceeded to see how they were and to my astonishment found that they had all—except two obviously unhealthy specimens—completely changed colour, being now quite pale instead of the normal deep black. Throughout the night they remained pale, but after dawn they gradually darkened until they had resumed the normal black. I naturally repeated this observation over and over again and found the nocturnal change of colour to be a normal occurrence.

Particularly beautiful were young Lepidosirens which in about a month had grown to a length of about 6 inches. Their day colouring was dark with round spots of yellow. At night they duly became pale except for the tip of the tail which retained its dark colour (Plate XXII (*a*)). On holding them close to the lamp one could see distinctly almost all their internal organs—intestine, liver, heart and so on. The eyes, hardly visible during the day, stood out clearly as dark beads. The dark pigment cells, forming during the day a densely crowded layer of greatly branched cells spread out parallel to the surface of the skin, their branches so intimately intermingled as to constitute a continuous light-proof coat, had now shrunk into tiny dots so as to have no effect on the general appearance. The yellow pigment cells however were unchanged, so that the young loalach in its night dress was nearly colourless except for round bright yellow spots.

In the full-grown fish the nocturnal change in colour was much less striking, the pigment cells being now hidden away under the greatly thickened epidermis.

Some years after the Lepidosiren Expedition Mr Pride brought to me in Glasgow a single specimen of Lepidosiren in which only the yellow pigment cells were present so that the whole fish was of a yellowish colour instead of the normal dark grey or black. This 'albino' aberration must obviously be of rare occurrence, as I never met with one among the many hundreds I examined in the Chaco, nor did I ever hear any rumour of it amongst the Indians.

A fascinating discovery in the embryology of the lungfish, in which I was forestalled by Semon in his investigation of *Ceratodus*, has to do with the teeth. Originally, during the early stages of their development, these are sharp and conical like the teeth of other fishes adapted for catching and holding their prey. As development goes on, however, the originally separate teeth become joined together into great ridged plates adapted rather for mastication, for crushing the food and not merely seizing it.

Those who devote themselves to research in departments of knowledge outside the 'exact' sciences concerned with things that can be measured or weighed, are every now and then impressed by the intrusion of factors which can only be described as mere chance. Such was the case during my Lepidosiren expedition: for example the extraordinary piece of luck that made my arrival in the Chaco practically synchronous with the onset of the—frequently quite irregular—rainy season and therefore with the commencement of the breeding season of the loalach.

Another case of similar good fortune in a matter of detail happened on 19th November when one of my Indian collectors brought in a lot of Lepidosiren larvae so crowded together that they were all dead when they arrived at the station. Not only so but they had become macerated in the warm water and many were in process of disintegration. As they looked interesting, I shook them up to aid the process, with the result that the skeletons became fully displayed by the disintegration of the soft parts.

Now it had happened, before I left England, that the Royal Physical Society of Edinburgh, of which I was a member, had been greatly interested by the discovery in Old Red Sandstone rocks of the North of Scotland of a strange little fossil to which the name *Palaeospondylus* had been given by Traquair, the greatest authority of his day upon fossil fish. Traquair studied and accurately described the structure of these fossil skeletons and there then emerged the question, what kind of a creature had this *Palaeospondylus* been in life? To which group of the animal kingdom did it belong? Traquair decided that its closest affinity was with the modern lampreys—a type of fish that is without the jaws that open and shut in the case of ordinary fishes and other vertebrate animals—and this decision was generally accepted. *Palaeospondylus* was an archaic kind of lamprey.

I had myself been particularly interested in *Palaeospondylus* and had in my memory a clear picture of its skeleton as described by Traquair.

What then was my astonishment to recognise in the macerated skeletons of the Lepidosiren larva a remarkable agreement with the skeleton of *Palaeospondylus*. The heavy lower jaw had dropped off and it was obvious that these macerated skeletons if embedded in mud and converted into fossils would closely resemble the fossilised skeleton of *Palaeospondylus*. The agreement was in fact so striking that there seemed no escape from the conclusion that *Palaeospondylus* was no lamprey but a primitive type of vertebrate closely allied to the modern lungfish. That this conclusion was correct will, I have no doubt, one day be generally accepted, when naturalists come to be more familiar with lungfish and their development.

Among the fish inhabitants of the Chaco, *Lepidosiren* stands supreme in scientific interest on account of two of its special characteristics. The first of these is its archaic character as a surviving representative of a relatively early stage in the evolutionary history of the Vertebrates which fact gives special weight to any evidence it may divulge regarding the evolutionary stages through which these animals and their various organs have passed.

The second rests on the discovery that emerged from the study of *Lepidosiren* that the cells of which its body is built up are of quite unusual size, as is also the case with the details of structure of the individual cell. Such details are of great scientific importance. For example, the chromosomes—rod-shaped structures which become recognisable at certain periods in the life history of the cell—have been proved to be the actual carriers of heritable qualities from parent to offspring. As a consequence detailed knowledge of the appearance and behaviour of these chromosomes is of the greatest importance to students of heredity.

Inspection of Plate XIX (*b*) which depicts, under the same scale of magnification, a particular stage of the reproductive cells of the male *Lepidosiren* for comparison with the corresponding cells from a mammal, brings out this difference in size of the chromosomes, and gives an indication of the great advantage given to the researcher into the physical basis of inheritance who has at his command material of *Lepidosiren* rather than that of man or other creature in which the chromosomes are much smllaer in size.

Apart from the lungfish the swamps and rivers of the Chaco possess a rich and varied fish fauna. Altogether I collected some seventy-five species—the majority belonging to two great groups—the Siluroids or catfish, and the Characinids. Other groups represented were the Chromides or Cichlidae and the Gymnotidae each with four species. Of the former *Acara tetramerus*, a little fish flattened from side to side, was often to be seen just under the surface of the water of the swamps and travelling along with a series of jumps into the air. A member of another group, the swamp eel *Symbranchus* has already been alluded to as having a breeding nest at the bottom of the swamp much like that of Lepidosiren. A fascinating discovery in this connexion was made later on by Agar. One of the most characteristic features of *Symbranchus* and its allies was supposed to be the complete absence of paired fins. Agar, however, was able for the first time to follow out the developmental stages of *Symbranchus* (his material was later worked out in detail by Sr Monica Taylor S.N.D.) and to his astonishment found the young larva (Plate XXIII (*a*)) to be provided with a pair of enormous pectoral fins richly supplied with blood and in life kept waving about—functioning as respiratory organs. When no longer required owing to the young Symbranchus becoming free from its confinement in the burrow these enlarged fins are simply shed and the adult shows no trace of their previous existence.

The Chaco Siluroids or catfish illustrate one of the chief peculiarities of this group of fishes, in which one misses the normal coating of scales—these thin overlapping plates of tough bony material so characteristic of ordinary fish. In one set of Siluroids these have become replaced by a strong armour of thick bony plates while in the other the surface of the body is completely naked except as regards the head and fins. While hard up for provisions on the Pilcomayo, we ate indiscriminately any fish we were able to catch and we found the naked Siluroids, called by our soldiers Bagre, were particularly suitable for roasting owing to their richness in fat. Some of these naked Siluroids are of enormous size and I found a native fishing in the River Paraguay, probably for Sorubin (*Pimelodus*) and using as bait whole oranges!

The armoured Siluroid *Callichthys* has peculiar breeding habits, the mass of spawn being attached to the underside of a kind of raft about a foot in diameter and composed of fragments of swamp plants buoyed up by a layer of foam containing bubbles of air underlying the raft and mass of eggs (Carter).

Amongst the Chaco Characinids is the palometa already mentioned in connexion with its almost razor-sharp teeth in the form of flattened blades which are used by the Natokoi Indians in lieu of scissors for cutting the hair.

Amongst the Chaco fishes I found two interesting interlopers far removed from the sea, their ancestral home. One of these was a Sole (*Solea jenynsi*) and the other a Sting-ray (*Taeniura dumerilü*) a relative of our skates but bearing a sharp serrated spine on the upper side of its tail near the base, and capable of inflicting a nasty wound. One that I speared in the Riacho Negro opposite Concepción dropped a number of nearly mature young—showing that the egg goes on with its development within the body of the mother instead of being laid in the protection of a horny egg shell as is the case with the ordinary skates.

One of the most interesting things in the natural history of the Chaco fishes has to do with the arrangements by which they are able to absorb the oxygen necessary for life at a time when the gills are no longer able to obtain it from the water that normally bathes their surface. Carter, who made a special investigation of the matter, found in different Chaco fishes three different types of arrangement whereby the necessary oxygen is absorbed directly from the air. The simplest of these occurred in *Hypopomus* or *Rhamphichthys*, one of the Gymnotidae, in which the gills, contained in an unusually large gill chamber, had their surface greatly increased by forming broad thin folds. This fish, when kept under the surface of water rich in oxygen, was able to get along quite well by breathing in ordinary piscine fashion, but on the other hand in water deficient in oxygen, it drowned if prevented from coming to the surface to fill its gill-chamber with air. In the two Siluroids, *Ancistrus* and *Callichthys*, air was swallowed, and the oxygen extracted from it by modified portions of the digestive tube—in the first mentioned the stomach, and in the second the intestine. Of the two, *Ancistrus* was found to be not entirely dependent upon air-breathing, being able to survive without it in well-oxygenated water, but *Callichthys*, on the other hand, drowned if kept under water however rich in oxygen.

Finally, in *Erythrinus*, one of the commonest of the Chaco Characinids, a portion of the air bladder, which in the normal modern fish acts as an automatic float, had reverted to its ancestral function when it was one of a pair of lungs. It still remained connected with the alimentary canal by a wide open tube equivalent to the windpipe of ordinary land animals, and was differentiated into three distinct portions, one behind the other, of which the middle one had a thick spongy wall full of cavities with lining richly supplied with blood and clearly constructed so as to function as an efficient breathing organ. *Erythrinus* again, though it constantly visited the surface to swallow air, was found to be not absolutely dependent upon doing this so long as the water was rich in oxygen.

The main object of J. S. Budgett in accompanying me on the Lepidosiren expedition was educational. He had planned to make the science of Zoology his life's work. He had completed the first part of his Tripos at Cambridge and he felt that the gap between it and the second part would be well utilised by the expedition to South America on which, while proceeding with the laboratory side of his work in studying the anatomy of animals, he could at the same time gain field experience of tropical nature. He felt, as I did, that the main task of the naturalist is to learn about living animals in their natural environment; and that the importance of laboratory knowledge of their anatomy lies mainly in its bearing upon their living activities and through them in the light it casts upon the means by which the evolution of the animal kingdom has been brought about.

Accordingly during our sojourn in the Chaco Budgett devoted his time, in part to dissecting and examining the various types of animal available in that region, in part to field natural history. At the same time I felt that he would greatly benefit by grappling with some particular problem of original research. The local Amphibia—frogs and toads—seemed to offer a particularly favourable subject for such research, the more so as the commencement of the rainy season meant just as in the case of the lungfish the commencement of their breeding season. Budgett took up this research with enthusiasm and his results are to be found recorded in a valuable memoir.*

An important preliminary part of his work was to track down the various performers in the marvellous amphibian concert that filled the air during the hours of darkness. Having identified the various species he next proceeded to investigate their reproductive habits, and finally he made an intensive study of the embryology of the species of which he could obtain adequate material.

The subjoined table summarises Budgett's analysis of the amphibian orchestra:

NAME	VOICE
Leptodactylus ocellatus	Regular repetition, beginning low note and ending on higher one; also like the drumming of snipe.
L. bufonius	Shrill, sharp 'ping' continually repeated until practically a continuous sound.
Atelopus nigricans	Two clear musical 'pings', followed by long descending 'trill' like that of a greenfinch.
Paludicola fuscomaculata	Mournful kitten-like cry.
Pseudis lamellum	Succession of sharp croaks like castanets.
Bufo marinus	Bell-like notes, middle higher in pitch.
B. granulosus	Continuous bell-like tinkle.
Phyllomedusa hypochondrialis	Sound like a dozen men breaking stones.
Hyla nana	High-pitched 'scraping' note.
H. phrynoderma	Like quacking of a duck.
H. nasica	Ditto, but lower pitch.

Apart from the human interest of identifying the members of this wonderful amphibian orchestra there arose important

* See *The Work of John Samuel Budgett*, Cambridge, 1907.

scientific problems connected with their existence in the peculiar environmental conditions of the Chaco. How, for example, did these creatures, still only imperfectly emancipated from the ancestral aquatic environment, cope with the periods of intense drought in which the presence of the water necessary for the tadpole stage of their life history was often restricted to short periods of a few hours following casual showers? Different species of Chaco frogs were found to solve this problem in ways differing in detail though possessing one common feature—namely, that the early stages of development are hurried through so as to occupy much less time than in the case of the frogs and toads of ordinary temperate climates which are without the violent alternation of damp and drought. Then the clear jelly which surrounds the eggs in ordinary frog spawn is in many Chaco species beaten up by the male during the act of spawning into a white froth which in some cases (*Paludicola*) floats on the surface of the pool, the young when they reach the tadpole stage wriggling down into the underlying water; while in other cases (*Engystoma*) the mass of froth is deposited in a burrow with a narrow opening excavated under a log in the forest; while in still others (*Phyllomedusa*) the female about to spawn climbs up, carrying the male on her back, on a plant overhanging a pool and deposits the spawn in a kind of funnel formed out of a single leaf or out of a group of leaves.

In each case the attainment of the tadpole stage is marked by the skin producing a kind of digestive juice which liquefies the jelly and allows the young tadpoles to drop down into the underlying water to go on with their development.

A feature of special interest in the natural history of the Chaco frogs and toads was the distinct relation between their habits and their desirability as prey. The Nup-mik-ting (*Leptodactylus ocellatus*), the commonest local frog, was also the favourite food of frog-eating animals such as fish, *Ceratophrys* and other large frogs and toads, and birds such as the Tum-um-hit (*Cariama*); and was noteworthy in its shyness, its skill in hiding itself and its agility in flight. And so also with other edible frogs. On the other hand, the uneatable toad Kilelik (*Bufo granulosus*), possessing large glands on each side of the head which produce an unpleasant secretion, was sluggish in its movements and made no attempt to escape. It was interesting to notice that in the young stage before the protective glands developed this toad had the habit of flattening itself out in the most extraordinary fashion, stretching out its four limbs perfectly stiff and rigid so as to make itself practically unswallowable.

I had naturally a special interest in the Amphibia as being the group most closely allied to the lungfish, and one of my special quests was for South American representatives of the more primitive Amphibia which, as in the newt but unlike the frogs and toads, retain the tail throughout life. No such tailed amphibian had so far been discovered south of Central America. As soon as my main task had been completed I set myself to make inquiries among the Indians as to whether they knew creatures agreeing with my description and pictures of tailed amphibians. At first the replies were all negative—then tales began to come in of a creature called 'lauhun'—froglike but with a well-developed tail. And finally the news of a lauhun being brought in by some Indians from the westward. At last it arrived but, alas, it was not what was wanted. It was, nevertheless, a specimen of great interest being the enormously overgrown tadpole of *Pseudis paradoxa* which, through non-development of its thyroid gland, had remained a Peter Pan among frogs, entirely undeveloped sexually, though reaching a size much greater than that of the normal adult.

Before returning to my diary something must be said about the Indians of the region where the lungfish work was carried out. They were ethnographically of minor interest as compared with the Natokoi of the Pilcomayo for they had not been isolated from contact with the white man. Missionaries of the South American Missionary Society had been working amongst them for some nine years and, apart from that, their territory extending to the River Paraguay facilitated trading with the Paraguayans. They belonged to the Mushcui race, as already indicated in Chapter VII, a branch of the Orejudos, so called from their wearing a disc of wood in the ear. The ear is pierced in boyhood, a small stick inserted in the puncture, and replaced from time to time by a flat disc of larger and larger diameter until in an old man it may be over 3 inches in diameter, the lobe of the ear being reduced to a narrow strand round the disc; producing a peculiarly repulsive appearance when the disc is removed.

In former times a similar disc was worn in the lower lip which was thereby caused to hang down in tongue-like fashion. Hence the name 'Lengua' (Sp. tongue) by which the Mushcui are still commonly known to the Paraguayans. In spite of their intercourse over a prolonged period with peoples of European race they remained in their general features typical red men, though somewhat more advanced in their civilisation than the Natokoi.

They had advanced from the stage of having only one headman of the tribe—they had now their witch-doctor as well as their 'cacique' or chief. Their personal names were descriptive nicknames, and it was a point of etiquette never to mention a man's name in his hearing. My own name I learned through the missionaries: it was *Oyankintimgyithlna-apkatik*—otherwise Pineapple head—for in those days I had a mop of rather upstanding hair. Budgett's was *Elyithwasé-abatung*—otherwise red mouth—for he had a red moustache and beard. The personal name is not necessarily permanent. If a man's son 'so-and-so' attracts attention by some outstanding occurrence he then comes to be known as 'the father of so-and-so'—just as sometimes happens in our ancient universities when outstanding prowess of Mr Snooks in the athletic field leads to some professor of world-wide reputation achieving at last a position of distinction in undergraduate circles as 'the father of Snooks'.

Place names also were descriptive—indicative of some characteristic of the place or recalling some event that had happened there.

Clothing and equipment were much the same as with the Natokoi. The garment of the men was normally more ample and woven of wool in correlation with the fact that the Mushcui possessed flocks of sheep. The garment of the women was formed of cierbo skin, dried and worked between the hands until soft and flexible. When in 'full dress' the face was painted red with urucu but without the black lines of the Natokoi. The Mushcui women were not tattooed as was the case with the Natokoi women.

Unlike the Natokoi the Mushcui had verbal expressions for numbers, apart from the Spanish uno, dos, etc. of which most knew at least the lower numbers. Using fingers like the Natokoi, beginning with the left hand: five—'one hand'; six—'arrived at the other hand'; ten—'two hands'; twenty—'hands and feet'.

As with the Natokoi religious beliefs were centred in evil beings, the cause of disease and other misfortunes. Medical treatment again consisted of incantation to drive away the evil spirit. It was of interest that these evil beings were called 'Askuk', a vague term corresponding rather with *bicho* in Spanish or bug in American. On the other hand, an Indian who lay ill with a touch of pneumonia related to me how during the night a 'tall' being had come up out of the swamp and touched him on the chest, which suggested rather the idea of a human form.

After death the Indian's body is bent into a sitting position and tied up in his garment. He is then carried off and buried and spiny caraguatá piled up on the grave. Where a mother dies her infant is buried alive with her. Children born after the death of the father are also killed. Apart from such cases there was, I gathered, a good deal of infanticide among the Orejudos, the infant being chosen and killed by the witch-doctor by a blow on the back of the neck with a digging stick. The missionaries had, I gathered, a hard struggle to put a stop to the practice among the Indians frequenting the Mission settlement.

As already mentioned the Orejudos though showing differences in many details of custom from the Natokoi were like them typical red men. They possessed the typical psychology of that race—a psychology well illustrated by a story of the Natokoi told me by Pride when on a visit to me some years after my sojourn with him in the Chaco.

He had been commissioned by the Paraguayan Government to endeavour to discover the fate of an explorer, Ibarreta, who had penetrated into the Pilcomayo region and never returned. Pride had traced the expedition to the country of my old friends on the Pilcomayo. The meeting with the Indians had been quite friendly and Ibarreta proceeded to photograph them. He arranged the group, putting his hand on the Cacique to guide him into a proper position. That night he and his men were killed. For a stranger to put his hand on the Chief is a thing that 'is not done'.

On hearing this story, entirely in accord with my knowledge of the Natokoi, I could not help recalling how different is the psychology of the black race. Our negroes on the Pilcomayo were most excellent members of the expedition. If they misbehaved they were 'walloped' like naughty children; they reformed their behaviour, and gave not the slightest trouble. Obviously the management of a team of red men on the one hand and of black on the other has to be along quite different lines to be successful.

Experience in later years served to drive home into my mind the crippling disability imposed upon parliamentary government in its dealings with other races by failure to appreciate such racial differences in psychology.

It has already been made abundantly plain that the Mushcui Indians gave me all-important aid in my Lepidosiren work. But my relations with them were of a quite different and far less intimate kind than those with the Natokoi of the Pilcomayo. Everything the Natokoi did for me was done 'for love' as the saying is. Nothing else, for example, would have induced Chimaki and his companions to risk their lives in accompanying me through the territory of the hostile Mushcui between Fortin Page and the Rio Paraguay. The help given to me by the Mushcui of Waikthlatingmayalwa on the other hand was given in return for payment, reinforced by the influence of the missionaries and also by the belief of the Indians—natural enough in view of my strange pursuits—that I was a particularly potent witch-doctor. It was at times desirable to encourage this belief. It unfortunately happened one day after I had taken a snapshot of a powerful cacique that within a few hours one of his eyes became greatly swollen and inflamed, and it seemed clear that I had shot an evil 'askūk' into the eye. Things looked rather awkward but as I walked casually away from the table where I had been working I was observed to stoop down and apparently pick up a little earth (actually anhydrous chromic acid which I had in my hand) and drop it into an enamelled basin of water (actually strong alcohol) which immediately burst into flame. I carefully refrained from looking back but I knew well how the talk would be of my power to burn up the water of the swamp with all its inhabitants on which they lived; and the consequent desirability of good behaviour!

Towards the end of January owing to the general flooding and increased depth of the swamp we decided to construct a dug-out canoe in which to navigate the deeper portions. We chose a large Palo Borracho (*Chorisia*), the soft timber being easy to excavate with axe and adze. A suitable one grew close by the station so we felled it and a couple of days work gave us a very serviceable looking vessel which, by means of palm rollers and with the assistance of a couple of bullocks, we launched triumphantly on 6th February.

10th February: After breakfast I started off with a couple of Indian boys in the canoe for a voyage through the swamp. Pride started with us as passenger, wishing to visit his island garden from which he had been cut off by the depth of the water. After disembarking Pride we turned Southward with the idea of crossing to an island away on the far side of the swamp. Until Pride left us the going was easy, following as it did the beaten track to the garden, but now we had to push our way through the thick and tangled vegetation. In places this was composed mainly of coarse aquatic grasses and there we could get along fairly well, but in others we had to traverse wide stretches of papyrus-like rushes, growing to a height of over six feet above the water surface and so close together that pushing the broad and rough-barked canoe through them was most laborious—I pushing from behind, up to the breast in water, while the two boys guided with poles. When we reached the middle deeper part of the swamp the rushes gave place again to grassy vegetation so progress became easier and I was able to get back into the canoe, and punt it along. At last, after traversing another difficult expanse of rushes, we reached the island and found it covered for the most part with palmar with some patches of monte. After exploring and doing some collecting we re-embarked, picked up Pride, and reached the station about 3.30.

It was a fine moonlight night so I determined to invade the recesses of the swamp in spite of the discouraging advice of the Indians regarding the evil spirits that were there abroad at night time. The air was perfectly still and the calm dark waters of the swamp, dull during the day with their monotonous vegetation, were now fields of great white night-flowering water lilies—a beautiful spectacle in the moonlight—the only drawback to its appreciation being the mosquitoes which were particularly bad.

As we pushed along I suddenly heard to one side the characteristic splash of a loalach. I could not see it, but Budgett who was in the bow of the canoe immediately asked what was the big white fish: could it be a Dorado (*Salminus*): it was 'as white as the lilies'. We continued on for a bit, but there was nothing further of interest.

By this date the level of the water was distinctly falling so I decided to make an attempt as soon as possible to get my precious material to Paraguay and, this accomplished, to have a short recuperative holiday before returning to investigate the dry season habits of the lungfish. So several days were devoted to the task of packing. I had brought with me from England a specially constructed laboratory box of teak, subdivided into many compartments and trays to accommodate bottles of chemical reagents, dissecting instruments, and so on. So, emptying this of its original contents, I packed into it the hundreds of specimen tubes containing the eggs and embryos, wrapping each tube in coarse muslin which I had brought specially for this purpose, fitting them all tightly into the various compartments so as to prevent them from shaking about. [The result of taking this care was that the box on being unpacked at Cambridge was found to have not a single tube broken or cracked.]

18*th February:* About 4 a.m. we were roused by the bell, and at once rose and packed our bedding. Coffee appeared, and after a hasty breakfast Budgett, Yamatenpekapkatik (Snailhead—the insulting nickname I had given to an Indian youth 'Tim'), and I started off on foot, the others remaining behind to accompany the cart. Following for a time a rough pathway we then struck across the palmar and joined the cart-track at a point where it cut through the monte and where the missionaries had constructed a gate to keep the cattle from straying. For some distance now the going was fairly good, there being comparatively little water on the track. Our way lay between islands of monte, while away on our right we could see the great swamp which we were soon to cross—really an extension of that by the Mission station. In the montes we shot some charátas while on the path were many animal trails—Tiger, Aguará guazú (*Canis jubatus*), Aguará mi (*Canis azarae*), and Aguará popé (*Procyon cancrivorus*). The path was often entirely hidden under the rank growth of vegetation and then we followed it in Indian fashion, feeling our way along it with toes turned inwards. At last as we approached the swamp the ground was everywhere covered with several inches of water and all trace of the track was lost. There came now about a mile and three-quarters of very heavy going through the swamp. The water was usually about knee-deep, but above it was a dense mass of grass tangled together with Convolvuli and other creepers and it was terribly hard work slowly toiling through this. However, a couple of hours brought us out of the swamp into the palmar beyond and here the water was only a few inches deep and the creepers less luxuriant. Rising out of the water everywhere were great ant-hills covered by a luxuriant growth of Asclepiads, Convolvuli, Caesalpinia etc. From one a water-rail flew up as I passed close by and peering in I saw a Rattlesnake no doubt just disappointed of its meal. Then again the going became more difficult; deeper water and the great heat very exhausting. At last after seven hours marching we arrived at our stopping place—the old mission station. Trousers and socks were completely worn through, legs much cut about, and worst of all my alpargatas—coarse canvas shoes with rope soles—had shrunk in the long soaking and my feet felt as if crushed to pulp and very painful.

After making a fire and having a meal of roast charáta we prepared to rest for the remainder of the day but alas! a messenger arrived to say that the cart had got stuck so we had wearily to retrace our steps for about a mile before we really pitched camp. A sharp southerly wind blew up and the night was delightfully cool with practically no mosquitoes.

The four following days were much the same, starting soon after daylight, heavy marching through flooded palmar and swamp, squeezing one's way through the thick matted rainy season's growth of vegetation, a halt when men and bullocks were completely done up, the rest of the day devoted to repairing the effects of the harsh cutting grass upon our garments—fitting them with new fronts of deerskin and replacing worn-out footgear with new moccasins, then towards evening a prowl round in search of game, an evening meal; and so to the night's repose upon the sodden ground, sheltered of course by the indispensable mosquito net strung up between two palms or bushes.

Of game there was always a sufficiency—an occasional Cierbo or Camp deer in the open country and Charátas and Mirikinás in the patches of monte.

21st February: A good sleep, the first for many nights. A cup of tea at first daylight and then onward for about a league when we stopped for the day having reached a fairly good camping spot. In the afternoon I went off to explore the neighbouring monte composed of Acacias, Chorisias, Guayacans etc. bearing many epiphytes—the white orchid *Brassavola*, the air-flowers *Tillandsia*, ferns and cacti. Of animal life there seemed almost nothing: not a bird or mammal: only trails of an acuti and a small armadillo. Rain poured heavily and I was drenched to the skin. As I write it occurs to me that I have not mentioned that waterproof garments—oilies or mackintoshes—are really *not* part of one's equipment in Chaco exploration! A thick flannel shirt is the only clothing of the upper part of one's body: the thickness is of importance as a protection not merely against cold but also against heat—on the one hand keeping out the radiant heat of the sun and on the other its moisture from perspiration giving an important cooling effect by evaporation—aiding in fact the normal action of the moist skin which counteracts the heat constantly produced in warm-blooded animals like man by the living activity of the body.

I was approaching the end of my time with the Mushcui Indians so I felt that discretion would now permit of my sounding them about my old friends the Natokoi, without of course revealing anything about my doings with them. I had merely a smattering of the Mushcui language so I put my questions through Hunt.

I first asked: Were the Suhin (=*Chuniapis*, *Choroti*, *Tapiete*) women tattooed? Yes, they were tattooed on nose and chin, while the Towothle (*Nimká*) women had their cheeks tattooed as well.

Then: Did they know people who did not wear ear-discs? Yes—the *Ai* or *Kanaktaa*. Did their women tattoo? Yes, they had tattoo marks on nose, chin and cheeks.

Were they good people? At this Poit hissed out *sŭmché* (=bad) with more expression than I had hitherto seen on a Mushcui countenance. They were *niptana* (tigers), very wild and they fought continuously with the Towothle.

Have they domestic animals? Yes, mules.

Where do they get them? By stealing from other peoples.

Where do they live? On the banks of a double river (obviously the Pilcomayo) about ten days' journey to the South-West.

Do they have clubs? Yes, large clubs with which they fight.

Is there anything peculiar about their bows? Yes, they have a string along the back, and their arrows are pointed not with iron but with cascarandá wood.

I had learned much about the various branches of Orejudo Indians from the Natokoi and it was interesting to hear about the Natokoi from the Mushcui. Incidentally, the conversation confirmed my discretion in keeping hidden my Natokoi associations, for they would undoubtedly have discouraged my so valuable Mushcui assistants in the Lepidosiren investigations.

22nd February: At daylight Budgett, Tim and I set off to hunt along the edge of a large monte to the Eastward of last night's camp—which by the way bore the Lengua name of 'the place where the young woman died who sang at night'. What exactly the story was and whether she really deserved her fate I could not find out.

We had first to pass through about a mile of deep swamp whose tangled vegetation soon reduced our trousers to shreds. Then coasting along the monte we shot a fine Ringtailed Cuati (*Nasua rufa*) in the branches of a Quebracho tree.

Then we were overtaken by an Indian with a note from Hunt announcing an accident to one of the yokes and asking us to return which we did after a fruitless further hunt of 3 miles or so along the monte edge. We were fairly tired by the time we reached the cart and were rather disgusted to find that the yoke had been repaired and all was ready to proceed. Budgett and I allowed the cart to go on ahead while we had a short rest and then followed on in the track of the cart—marvellously easy going as compared with our usual struggling through the unbroken vegetation. Passing the old corral where we had spent a siesta on the outward journey (28th October) we traversed the first of a series of picadas (forest tracks) and then had to stop, the picada in front being impassably choked with two years' growth of vegetation. The place where we now encamped bore the unpleasant name Thlagna simpehē or 'the place of horseflies' and it did not belie its name. It was late when we camped and we hurriedly prepared our beds as much lightning away in the north seemed to presage a storm. I had fixed my bed close to the edge of the monte and had not lain down for more than half an hour when I was roused by the snap of first one then another dry twig—the infallible sign of a large cat, no doubt a jaguar, close by. However, our visitor's reconnaissance did not develop further though the cattle showed much restlessness.

23rd February: The early part of the day was spent in clearing a way through the range of picadas extending for about a mile and a half and our day's journey was short, only about three leagues, bringing us towards sundown to the margin of a deep swamp stream—the same that we had crossed already several times—at a point known as Yowe ohuk kamūk (=large based Kamūk) from the growth of *Thalia* (Kamūk).

24th February: We got the carts unloaded while Hunt, Sibeth and the Indians set to work to cut a road to the creek through the quite impassable growth of papyrus-like giant rushes.

I waded through the stream and went off to shoot charátas in the monte. On my return I enjoyed paddling about in the cool deep water while Hunt and Sibeth got the carts across, four bullocks yoked to each. In the deep central part the bullocks had to swim, Hunt and Sibeth on either side to direct their course. Meanwhile the various packages were carried across by the Indians on their shoulders [Plate XXIII (*b*)] and I was horror stricken to see that one of my panniers containing two large jars of specimens was being carried on its side so as to risk the loss of spirit from its precious contents. Happily the leakage was slight and no real damage was done. After reloading the carts and refreshing ourselves with a cup of tea we proceeded on our way and after about 2½ leagues mostly through swamp we arrived at Yatiktuma ŭptapithmūk nŭpnŭksĕhē (= the armadillo with yellow fat) where Budgett and I had camped on the outward journey on 27th October.

The breeze was westerly so the Indian Tim who had gone ahead to shoot charátas set fire to the grass to clear the way for next day's journey. Unfortunately the wind suddenly changed round to the eastward and the vegetation being extremely dry we soon had an immense conflagration bearing down on our encampment and threatening the destruction of everything including our priceless collections. I felt that the only possible safety lay in burning lines of grass one behind the other, keeping each under strict control and in this way to form a barrier of burnt ground broad enough to stop the main fire. By exerting ourselves to the utmost we were successful, though for a time things looked critical when the main fire broke through a weak point in our line of defence. However, we were able by using wet garments, rugs and anything else handy, to beat out the flames when they had reached within twenty yards of the carts. From this away to the eastward stretched a blackened waste, swept clear of all vegetation—while away to the South-West we could still see the conflagration blazing freely until a late hour.

25th February: Started about sunrise and almost immediately began to traverse a swamp half a mile or so broad and waist-deep in the middle so that the water just entered into the carts. A short stretch of dry ground and then swamp again waist-deep all through and with such dense tangled vegetation as to make progress very difficult. After this, dry ground for a mile or two with fine montes and open grassland between. We set fire to the grass everywhere and it blazed famously preparing for growth of fresh young grass for our return journey. Then a range of picadas where the track bordered a deep swamp stream. The day was extremely hot with trying north wind so after passing the picadas we came to a halt for siesta, planning to do the couple of leagues to the Riacho Verde in the afternoon.

About 3.30 Budgett, Tim and I pushed on ahead through fine open grasslands between patches of forest, the latter now richer and more luxuriant, marking our approach to the Paraguay river —with fine Pindós (*Cocos australis*) in addition to the two common palms we were accustomed to farther Westward.

We soon reached the Riacho Verde and skirted along its margin for a mile or so. The water was still very high though it had fallen some five feet from what it had been when at its highest. The waters obviously teemed with fish for there was a magnificent show of fish-eating birds—White egrets (hundreds grouped together), a chajá or two, herons of several kinds, bitterns, innumerable cormorants, kingfishers. The Egrets looked very fine, their snowy plumage rosy in the rays of sunset and standing out conspicuously against the dark background of foliage.

After arriving at the crossing point we investigated the channel and found about 50–60 yards out of our depth. The carts arrived about sunset and we camped for the night on the soft muddy ground, the air permeated by a horrid smell of decaying vegetation. Soon after we encamped a large flock of Jabirús flew over and Tim brought one down with my gun. Cormorants were flying about in all directions and the Indians had a great time trying, unsuccessfully, to kill some—crouching down in the grass and sending stout sticks about a couple of feet in length whirling through the air as the birds passed overhead. They all seemed greatly to enjoy the sport in spite of its being without result. Cooking our supper was a difficult job as there was no fuel near at hand except palm stems which had been soaking in water during the whole rainy season.

26th February: A heavy day's work passing the river. Budgett and I constructed a small raft of five ten-foot palms lashed together and stiffened with a cross piece at each end. On this we lashed a wooden packing case into which we packed cameras and other small things to float them across the river, Budgett swimming in front while I, a non-swimmer, pushed from behind so as to help as much as I could. The carts minus their contents were brought to the water's edge, their flotation power increased by lashing half a dozen palms beneath their floor, and then hauled across by the Indians on the far bank until the water became shallow enough to yoke a couple of oxen. A larger raft of palm stems with a raised platform enabled our various packages to be ferried across in four or five journeys. The day was very hot and it had been pleasant to work in the cool water but we had no sooner finished our job than it clouded over and torrential rain fell with much thunder. Both Budgett and Graham ill with fever in the evening.

28th February: Yesterday's journey was through country obviously drying up, the track in many places not submerged but composed of soft and sticky mud which made walking most laborious. We camped early, the oxen completely done up with the heat. I roused the others at 3 a.m. and got everything ready to be off at daylight—Budgett, Tim and I as usual going on ahead, attacked by innumerable tabanos (horse-flies), dozens buzzing round our heads and biting whenever they got the chance. Up to a short picada near our third camp on the outward journey (26th October) there was still water standing in every little depression, but now things changed completely and the country became dry and parched. On reaching the spot where we had planned to camp we found the swamp where we expected to find water completely dried up, so we had to push on to a water course a few miles farther on though the oxen seemed to be at about their limit, with heads down and tongues hanging out. But disappointment awaited us again for the river channel was quite dry and hard. So we had again to push on and we were glad indeed when at last we saw ahead of us in the distance a number of Jabirús and Wood Ibises betokening the presence of water. The Oxen too seemed to shed their fatigue, realizing by their mysterious sense that there was water ahead.

Presently we reached what had been a deep rivulet on the outward journey and found in its bed a muddy puddle, a moving mass of fish concentrated in it, and with this we had to be content. The water was of the consistency of thick chocolate; not a drop could the pump force through my Berkefeld filter; but yet Graham managed to prepare an acceptable meal with the aid of some Manioc, Sweet potatoes and Pumpkins which we had got from an Indian tolderia en route. This we polished off 'con mucho gusto' in spite of the air being heavy with the odour of the decaying fish in neighbouring parts of the water course. Having satisfied our own needs we turned the Oxen loose on what remained of the puddle.

1st March: A weary tramp over country parched and dry and in many places burnt: not a spot of green visible apart from the patches of forest. However, at last we saw in the distance the hazy looking high ground of Paraguay and after two or three further miles of weary tramp, we were on the bank of the Paraguay river taking in great draughts of its delicious water and watching it curling and eddying away southwards and linking us up again to the outer world of civilization.

After an impatient wait of four days at Carayá Vuelta in the vain hope of being picked up by a passing steamer, we borrowed a canoe from the neighbouring estancia and despatched Sibeth and a couple of Indians with instructions to go downstream to Villa Concepción and arrange for the little steamer *Coco* to come up and fetch us. This she duly did; our precious cases were safely transported to Villa Concepción and placed in store; and the three of us—Budgett, Hunt and myself, took the next steamer down to Asunción.

CHAPTER X

A PARAGUAYAN INTERLUDE

14th March: We left Concepción at 9.20 a.m. in the *Aurora* and it was a delightful sensation to be carried along without having to use one's own legs. The river, though it had fallen some eight feet, was still high and the jacarés were still away in the interior—to return to the Paraguay later on as the esteros and lagunas dried up.

17th March: After a pause in Asunción and a pleasant dinner at the Stewarts' we entrained for Villa Rica, i.e. Budgett, myself and Mr Hunt who was to share our little holiday in Paraguay. At the first stop refreshments were to be had: a row of cows tethered close by, ready to be milked according to the demand. The undulating countryside, dotted with palms, was obviously suffering from prolonged drought though here and there a deep pool still contained a little water with Pontederia floating on the surface. A thick haze interfered with the distant view but we had glimpses of the Laguna Ipacaray extending for several miles on our left. Presently it was succeeded by low hills, apparently volcanic in their nature and richly wooded up to their summits. Between 10 and 11 we paused for 40 minutes at Paraguari for breakfast and thereafter our journey lay through parched country—often grassy expanses—until about 2.30 we arrived at Villa Rica—little changed since my visit of six years ago except that an ugly embankment now dammed the waterpool, then so picturesque with its white clad women doing their washing along its margin.

18th March: An annoying experience! Last evening when visiting the local barber I observed hanging on his wall a remarkable fossil—obviously the skeleton of a creature of special scientific interest as its features showed a mixture of Amphibian and Reptilian characteristics. At dinner I recounted enthusiastically my discovery to Budgett and said I proposed to do my best to induce the barber to sell. Besides ourselves, there was only one other guest dining —a German commercial traveller.

So immediately after morning coffee I returned to the barber's to open negotiations, but alas! to his poignant regret he had been induced to part with it to a German Señor who had just departed!

After this painful experience we proceeded to the market place which unlike that at Asunción is in the open air—a busy and picturesque scene, the women with their white garments—varied only by the shawl over their heads being sometimes black instead of white—squatted on the grass with their little selection of wares beside them—bananas, mandioca, dulce de leche, chipa, bundles of firewood, roughly modelled clay pipes, etc. The better looking ladies were very coy—turning away or hiding their faces when the camera pointed at them.

In the afternoon we studied the art of making ñandutí, the charming spider-web lace of Paraguay, worked with needle and thread on coarsely woven cloth stretched on a wooden frame about 18 inches square. On this cloth the pattern is sketched with pencil as a preliminary and after completion it is cut away. Our young instructress seemed to fit in well with the lines of U. E. Montes:

La Tejedora de Ñandutí

Graciosa, esbelta, pura y sencilla,
Con aleteos de mainumbí,
Al brazo lleva su canastilla
La tejedora de Ñandutí.

Flores de ceibo su boca imita,
Su talle es fino como el piri
¿Quién la resiste, si es tan bonita?
¿Y hace tejedos de Ñandutí?

Cárlos la adora, y oye en el sueño
Dulces palabras en guaraní,
Y que le llama su amado dueño
La tejedora de Ñandutí.

Ayer la dijo: Qué hermosa eres!
Oh Paraguaya, muero por tí!
Juntos haremos, si tú me quieres,
Muchos tejedos de Ñandutí.

Gracias, responde, pues soy dichosa
En las riberas del Tacuarí,
Donde es amada como una diosa
La tejedora de Ñandutí.

Mi novio cuida sus lindas cabras
Siembra mandioca, planta maní;
Más primorosas son sus palabras
Que mis tejedos de Ñandutí.

En su canoa me lleva al lado,
Me da fragrante peripotí,
Si lo supieras! le tengo atado
Con suaves lazos de Ñandutí.

Quién es más noble, quién es más rico
Que mi adorado? Feliz de mí!
Y coqueteaba con su abanico
Lleno de estrellas de Ñandutí.

Cogió, sonriendo, su canastilla
Y, con la gracia de mainumbí
Siguió su ruta, tierna y sencilla
La tejedora de Ñandutí.

The Weaver of Ñandutí

Gracious, graceful, pure and simple,
With winglets of mainumbí [humming bird]
In her arms she bears her little basket
The weaver of Ñandutí.

Flowers of Ceibo her mouth imitates,
Her figure is fine as the reed.
Who shall resist her if she is so lovely
The weaver of Ñandutí?

Carlos adores her and hears in his dreams
Dulcet accents in guaraní
And that she calls him her beloved master,
The weaver of Ñandutí.

Yesterday I told her: How beautiful you are,
Oh Paraguaya I die for thee.
Jointly we'll make, if you desire me,
Many a fabric of Ñandutí.

Thank you, she says, I am betrothed
In the district of Tacuarí,
Where is loved as a Goddess
The weaver of Ñandutí.

My sweetheart tends his fair goats,
Sows mandioca, plants maní,
More beautiful are his words
Than my fabrics of Ñandutí.

In his canoe he bears me at his side
Gives me the fragrant peripotí,
If he only knew it I have him tied
In tender bonds of Ñandutí.

Who is more noble, who is more rich,
Than my adored one? Happy me!
And she flirted her fan
Filled with stars of Ñandutí.

Smiling she gripped her basket
And with the grace of the humming bird
Went on her way, tender and simple,
The weaver of Ñandutí.

Among the English-speaking people we met in Villa Rica a main subject of conversation was the collapse of the great socialistic experiment of 'New Australia'—an aftermath of the disastrous general strike of 1890–92 when some 300 embittered victims, inspired by the eloquent preaching of socialist ideals by a journalist named Lang, determined to shake the dust of Australia off their feet and found a socialistic colony elsewhere. They found an ideal location for their experiment when the government of Paraguay offered to grant them a tract of country nearly six hundred square miles in extent on the banks of the Tebicuarí, including some of the finest land in the country, well watered, with rich pasture and luxuriant forest. Not only were the conditions extraordinarily favourable but the human material was of the most suitable type, not mere town-dwellers entirely unsuited as colonists but many of them trained countrymen—stock-workers, sheep shearers, country artisans. Each colonist contributed the whole of his means—varying from £60 to £1000—to the common fund.

In spite of the marvellously favourable auspices of its start the scheme was doomed from the commencement by the very human failings of the colonists. As one of them said to me 'Everyone wanted to be schoolmaster, no one to be scavenger'. Another—'Lane does the thinking and the colonists do the work'. Those who had brought musical instruments with them found their use more congenial than that of axe or saw or plough. Enthusiasm for looking after communal property soon died away: valuable agricultural implements and tools were left lying about in the rain uncared for and either lost or rendered useless. Finance got into a hopeless muddle. A strong and industrious worker was given the same pay as a lazy 'weary Willie'.

The one activity that never failed was the holding of meetings with interminable debates.

The end of the story was that the New Australians with their hardly won practical experience harked back to honest individualism. Some took jobs on railway construction and other work, and with their savings settled down, built houses, fenced in plots of land, made gardens and eventually became prosperous citizens with livestock, cattle, pigs, poultry of their own.

Though there was little enough new and unexpected in the tale of New Australia it was interesting to hear it direct from the lips of actual participants.

From Villa Rica we had decided to make a journey to the little town of Caa guazú (Great Forest) away to the northward, and accordingly saddled up with first daylight upon March 26th and left Villa Rica by the North road. For the first four leagues or so the country was open and undulating with considerable stretches of flat. Away to the East lay the chain of the Sierra wooded up to its summits. After about five leagues the thickly dotted ranchos indicated our approach to the river Tebicuarí—a fine stream but now so reduced by the prolonged drought that we were able to ford it. Pushing on after our mid-day meal we traversed a long and dreary stretch of open camp, the only incidents being our meeting a troop of several hundred horses; and a family group transporting cotton, carrying heavy bales on their heads while three pack-horses were also heavily laden. At last we reached the entrance of the seven league picada through the great forest and camped for the night: bitterly cold but managed to get a few winks of sleep, lying close to the fire and turning over so as to heat alternately now one side, now the other.

27th March: Soon after sunrise we got under way on our memorable ride through the forest, here much more luxuriant than we had been accustomed to in the Chaco. The majority of the trees were not large but every here and there was a veritable giant, somewhat slender in growth but towering up to a height of well over a hundred feet in the effort to reach the upper daylight, and usually with clean round bole, only occasionally showing the vertical buttresses familiar in descriptions of tropical forest. A good many palms and innumerable ferns—including many tree-ferns—with lianas in festoons between the trees: in correlation perhaps with the dimmer light the undergrowth was much less dense than in the montes of the Chaco.

The soil was coarse red sand with outcrops of red sandstone, and the road was merely a rough track, the wheel ruts several feet deep and often one of them three or four feet below the other, while the litter of broken wheels and axles testified to the severity of the test it gave to the bullock carts of the Yerbateros who provide most of the traffic. We passed several parties en route repairing damages—making fresh axles of tough Lapacho and binding them to the body of the cart with strips of raw hide.

Occasionally a little rivulet crossed the track, running sometimes in a steep-sided gorge, at other times in an open valley. In these sunlit gaps in the forest we had a marvellous display of butterflies and I cannot do better than transcribe from Budgett's diary as quoted in his biographical memoir by Shipley:

'But when we came to the little streamlets in the gullies, at first I could not believe my own eyes, for here and there were great flocks of butterflies sitting about over the rocks in masses of colour according to their species. There would be perhaps a patch two yards square literally covered with a large yellow butterfly, something like a Brimstone, only twice as large, the males being of a bright orange colour. Then a yard or two away there is another patch perhaps more compact, composed of a fine species of a rich brown colour with a double bar of white and orange, stretching right across the four wings and the body. A little further there would be a patch less densely but uniformly populated with a variety of species, from a brilliant little fellow, red and blue above with concentric circles of sulphur below and upon a black ground, to the huge *Morpho* with his blue wings measuring six inches from tip to tip. The large sulphur patches were perhaps more frequent. Then as one looked down into the gully, besides this Lepidopteran carpet of idleness covering the rocks, the air would be alive with more busily engaged individuals, partly of those already mentioned, but certainly also of the flashy energetic Heliconidae with their brightly striped wings of black and red or orange: their mimics the Danaidae (Ithomiidae) were also there. But when I walked my horse into the stream, and dismounted to let him drink, the air seemed filled with the flapping of flimsy wings which gently fanned my face, and, as I waited a moment or so, they began to resume their basking, the social species collecting in their flocks, while the more varied assemblages settled everywhere. Twenty-five were perched upon me at one moment the smallest measuring more than two inches across the wings, six were up on my gun-barrel, my horse too was pretty well covered, while a great *Morpho* had alighted on his forelock.'

In the dense forest itself we saw hardly a living thing—a Cuati and an Agouti, a pair of Trogons sleepily perched on a branch, and now and then we heard the shrill screams of the long-tailed Cuckoo, *Piaya cayana.*

After reaching the open we had a league or rather more to traverse before we saw on the far side of the broad valley of the I Hu (= Rio negro or Black Water), the houses of Caa guazú.

The village we found to consist simply of a gigantic plaza with a line of small thatched houses all round. In the middle was the Church and close by it the belfry—a curious watch-towerlike structure. All the inhabitants naturally turned out to see the strange visitors. We asked for the Fonda de Manfredini where we were to pass the night and found it on the far side of the square—a boliche or pub of a very primitive kind. All the chief inhabitants were congregated there. One seemed the local savant and I overheard his instructive remarks upon European languages 'It is difficult to read English or German for they do not represent sounds by letters as we do. For example the sound "ā" is not represented by "a" but by a "c", a "j" or a "k".'

After siesta, following a quite fair meal though with deplorable wine, I went off to reconnoitre to the North of the town while Budgett sketched. In the evening various citizens called to enquire our charges for taking photographs and seemed disappointed at our not accepting professional engagements.

28th March: By the time we reached the entrance to the picada on our return journey the sun was getting well up and it was hot enough to make us appreciate the delightful coolness and deep shade of the forest. About 12.30 we reached the end of the picada and camped by a patch of monte a little farther on. We were quite discontented with the water supply—a little runlet of tepid muddy water in a wheel rut—though we should have been glad of it at times in the Chaco. After breakfasting off a little stew of beef, rice and vermicelli I walked over to a group of Yerbateros returning from the Yerbales with horses laden with packs of Yerba mate and had a pleasant little crack over a 'mate á la bombilla'.

About 3 o'clock we moved on and by sundown reached an excellent camping spot on the banks of the Tebicuarí. We had a magnificent fire and cooked our evening meal, the horses tethered close by. We sat up late, Budgett and I becoming involved in an acrimonious discussion as to the position of horses' legs in cantering.

29th March: I decided to push on ahead so got up at 4 a.m. and had quite a fight to get my mule bitted and saddled as she didn't seem to approve of the idea of starting off alone. There was just enough light to distinguish the track and at daybreak I reached a beautiful little streamlet where I cinched up and had a drink of cold water to serve as breakfast. The mule would not go well—absolutely refusing to canter freely—probably from lack of confidence. At one point she did come a cropper but I was riding with the usual loose seat and feet loose in the stirrups so came off over her head and landed on my feet. Numerous wild guinea-pigs darted about among the tufts of grass. For the last couple of leagues to Villa Rica Mula went gaily, without a touch of the spur, and I arrived at the Villa just in time for morning coffee.

Next day I trained in to Asunción and spent a week there awaiting the steamer *Aurora* for the trip to Villa Concepción where I arrived on 9th April.

CHAPTER XI

CONCLUSION

14th April: In the morning I crossed over to the Mission station on the western side of the river and from there we made a start upstream in the early afternoon in a large canoe—deeply laden, for in addition to about half a ton of stores we had on board besides myself, Graham and two new recruits to the Mission—Hawtrey and Mark, two Messrs. Insley, and four Indians. Even with four oars our progress was slow and we made only a couple of leagues or so before stopping for the night.

15th April: We had only gone a short distance when I missed my large hunting knife and we put back to recover it as I assumed the Indians had missed it when loading up my gear in the canoe. There was, however, no trace of it and I was naturally vexed at such a serious loss.

[It was not until several months later that one day in Cambridge a package arrived by post containing my knife, somewhat rusty and with the sewing of the sheath all rotted away, and a letter from Pride explaining how after the waters of the Paraguay had gone down an Indian wandering along its margin had found the knife, and knowing the story of my loss had handed it to one of the missionaries. No doubt it had been carelessly dropped into the river by one of the Indians loading up the canoe and he had refrained from drawing attention to the fact.

The finding of my knife, the fact that the finder knew its ownership and that he was honest enough to hand over an article of such value, involved a set of striking coincidences. What made the recovery of my knife the more remarkable was that I had a similar lucky coincidence in recovering its predecessor on the Pilcomayo after it had been dropped unnoticed in the open palmar.]

On the way upstream we called in at a Brazilian coffee planter's and bought a goat and chicken. He also gave us the best coffee I had ever tasted, with rich fat upon the surface and a most delicious flavour. He said it came from a single quite phenomenal bush whose berries he kept entirely for his own household. Looking back from these days when so much is known about the propagation and improvement of plants of economic value it seems quite a tragedy to think of the lost opportunities centred round that coffee bush.

Less pleasant than our call at the coffee planter's was the onset of my worst bout of malaria on either of my expeditions. Lying on rough shingle by the riverside with a temperature of 104°, severe headache and violent sickness, that first night was the most unpleasant, the actual suffering becoming less in subsequent stages with increasing weakness.

16th April: Very ill all day but my good comrades rigged up my mosquito net on top of the cargo in the canoe and there I lay till our arrival about 1 a.m. at Carayá Vuelta, where I got ashore and lay in the missionaries' old hut for the next four days. Budgett arrived on the 18th and installed himself as a most admirable nurse. On the 20th Grubb and Pride arrived from Waikthlatingmayalwa and I retreated down river with them to Concepción, lying in the bottom of the canoe. A few days at Concepción and I was sufficiently recovered to make a fresh start upstream with Grubb and Hay on April 25th. Two days later we left Carayá Vuelta with fresh horses and did the journey to the station in three days, the country being now all dry and easily passable.

30th April: Budgett and I accompanied by several Indians went off to explore the swamp under its incipient dry season conditions. Right away to Pride's garden island it was now completely dry: further on there was a stretch with a few inches of water where there had been five feet. Some of the lungfish had retired into their dry season burrows—'etsasa' or deep nest as the Mushcui call it in contradistinction to the shallow burrow—'pukthlanmu' —in which the eggs are laid. We found lots of the latter, still containing water; sometimes by noticing the apertures just like a rat's burrow, sometimes by an Indian probing with his spear in a likely spot. When found an Indian would explore with his foot and the loalach if still in the nest would be pulled out; not too easy on account of the extraordinarily slippery slime covering the body. This slime by the way I had found to have a remarkable power of precipitating mud. We had been accustomed to use the regular travellers' dodge of clearing muddy water for drinking by the addition of a little alum but I found the same purpose could be served by keeping a live lungfish in the water—the mud being all precipitated and the water left quite clear.

As we crossed the island each Indian had cut himself a stout stick a couple of feet or so in length and sharpened at one end to form a digging implement and with these numerous burrows were unroofed so as to display their extent. In one case such a burrow, somewhat curved in plan, extended for a length of five feet.

Many of the lungfish were still, as already mentioned, in their breeding nests, but some had already begun to subside into the mud, usually by the side of a rooting tuft of vegetation; rush, bulrush or the ordinary coarse swamp grass.

During their active life when free in the swamp the lungfish had fed voraciously, storing up a large quantity of reserve nourishment in the form of fat deposited among the muscles especially of the tail. With the disappearance of the waters their food supply comes to an end and subsistence then depends entirely on this stored up nourishment until the advent of the next wet season.

4th May: During these last few days I have been getting many specimens of the charming little Opossum mouse *Marmosa pusilla*, called by the Mushcui 'Kilyabuktik' from Kilyoa—large, and abuktik—eye [already mentioned in Chapter VI, p. 85].

6th May: Today with Budgett's assistance I attempted to make coloured injections of the arteries of Lepidosiren to provide material for dissection later. We first replaced the blood in the vessels by injecting normal saline solution and then ran in coloured gelatin under about four feet of pressure. There was some difficulty owing to the fine branches of the arteries closing up in response to the irritation caused by the strange injection material. However, I had foreseen this possibility and had brought a supply of nitrite of amyl which physicians use to dilate the arterioles in cases of angina pectoris. I arranged a T-shaped tube in the mouth of the lungfish, leaving one branch free for the animal to breathe through and plugging the other with a little cotton wadding containing a glass ampulla of amyl nitrite. When the injection ceased to flow I broke the ampulla and closed the other end of the cross piece of the T so that the vapour of the amyl nitrite was inhaled; the fine branches of the arteries at once responded and the injection passed freely.

During our last days at Waikthlatingmayalwa much work had to be done; making and soldering up tanks and tubes of zinc and tin in which to transport the fully grown Lepidosirens, and then constructing wooden cases for their protection. Tin plate was in some respects more suitable but in the case of formalin specimens the oxidation caused rust staining which ruined their appearance, while the oxide produced in similar circumstances in the case of zinc was without this unpleasant staining effect.

I was also able to make interesting additional observations upon some adult lungfish placed in a deep pool dug in front of the station. One set of observations dealt with the time interval between successive intakes of air into the lungs. These turned out to be very irregular. Rough records of four different individuals read for example as follows, in minutes:

5, 2, 3, 3, 3, 4, 4; 3½, 8, 2, 1½, 3½, 8½;
4, 4, 4½, 5, 4¼, 4¼, 4⅓, 4⅓;
and 4, 4, 6, 3, 1, 3, 4.

As the day of our departure approached we drained the water from the pool so as to observe the behaviour of the lungfish when taking up his dry season residence underground. After the water had drained away the location of a loalach was marked by a depression two or three inches in diameter filled with liquid mud. Carefully watching this one presently saw the surface of the mud slightly raised in a dome-shaped manner, then a small opening appeared in one side of the dome and this increased to a triangular aperture about half an inch in length—obviously formed by the lips of the Lepidosiren moving apart. The loalach was in fact going through the process of breathing air as it does normally when free in the swamp, pushing the tip of its snout just through the surface and then inhaling air through the mouth opening. Next day—some thirty hours later—the loalach was in the same position with his head just beneath the surface of the mud, raising this into a rounded hillock near the base of which was a flattened triangular opening now permanently open owing to the stiffening of the drying mud. Through this opening the loalach was breathing regularly and continuously, causing a slow rhythmic heaving of the hillock.

To my great disgust a heavy rainstorm brought these observations to an end but my missionary friend Hunt was kind enough to continue them after our departure. He dug up many loalachs during the dry season and stated the burrow to average 2½ to 3 inches in diameter and from 12 to 20 inches in length. The opening in the plug of dried mud at the outer end of the burrow he found to be commonly subdivided into two or three rounded holes 'each a little larger than a peppercorn'. In the deeper burrows he sometimes found several plugs four or five inches apart. The deep end of the burrows was somewhat dilated and in this part lay the loalach with the tip of its tail folded back over its face. The narrow space between the surface of the loalach and the wall of this dilated part of the burrow was filled with slimy mucus secreted by the skin. As was to be expected the dry season loalachs taken from the dark interior of their burrows were light in colour but soon became dark under the influence of daylight.

Our stay at Waikthlatingmayalwa was now at an end. The Mission Station, only a few feet above the general level of the swamp, was situated on an island which became steadily smaller and smaller as the floodwaters rose during the rainy season up to about the end of January.

During this period the resident staff of the Mission consisted of Andrew Pride, R. J. Hunt, and Graham the recruit whose arrival synchronised with that of Budgett and myself. In addition to the missionaries were the two German helpers—Sibeth, stockman; and Wilhelm, cook.

I had been in my early days rather unenthusiastic about mission work among alien races: the task of destroying an existing religious faith seemed so much less formidable than the installing of another in its place. However, as Budgett and I witnessed the strenuous and selfless day-to-day lives of Pride and Hunt isolated from so much of the luxury and comfort of life in a civilised society any tendency to criticise became replaced by a feeling of deep admiration. Apart from more strictly spiritual teaching Pride occupied himself busily in instructing the Indians in the ordinary arts and crafts of civilisation—simple carpentry and house construction; the care of stock—cattle, horses, sheep and goats; gardening: while Hunt, in addition to assisting Pride, devoted much time to the study of the Mushcui language and the rendering into it of passages of Christian scripture. Greatest of all in the work of these missionaries was the diffusion among the Indians of the Christian virtues exemplified by their own clean and eminently Christian lives.

During our later stay at the station, after the Paraguayan interlude, the mission community was increased by the return of Grubb, the original founder of the Chaco Mission, from his furlough in Europe, accompanied by the two new recruits, Hawtrey and Marks, so that along with the varying number of Indian adherents the settlement now counted some twenty to thirty inhabitants (Plate XXIV).

An awkward complication which I have not mentioned in my

diary was the appearance early in January of an epidemic of the strange disease Mal de Caderas (Hip disease) which swept away with a single exception all the horses of the neighbourhood. The early symptoms of the disease were peculiar—a kind of locomotor ataxia—the hind legs apparently proceeding to walk away in a different direction from the fore legs. In later developments the hind legs became completely paralysed so that the animal could not rise from the ground. It became much emaciated, sores developed on its back, and finally death took place after extreme suffering. On 4th January I found my favourite little Chestnut lying away by himself in the monte, biting the ground in his agony and looking to me in the most pitiful way, and there was nothing to be done but to put an end to his misery.

We found that the Tapirs and Carpinchos were also dying off and when the dogs about the station did so too we began to wonder whether man too might not prove to be susceptible!

Later developments of pathological science have shown that mal de caderas is a form of trypanosomiasis caused by a blood parasite (*Trypanosoma equinum*) allied to that which causes the Tsetse-fly disease, so fatal to live stock in the 'fly belts' of tropical Africa. Such microbes are now believed to be normal parasites in particular species or races of wild animal which have in the course of ages become tolerant of their presence in the blood so that there is no disturbance to health. Such animals however may serve as 'carriers' from which infected blood may be transferred by bloodsucking insects to animals of other species or races which, unaccustomed to the particular species of microbe, have not had the opportunity of developing tolerance towards it, and accordingly react to its presence by developing disease. A well-known example of such an occurrence is the epidemic of sleeping sickness which invaded Uganda half a century ago and carried off many thousands of its inhabitants, due apparently to natives from the West Coast 'tolerant' of the particular trypanosome (*T. gambiense*), carrying it up the Congo river to be distributed by biting flies (*Glossina*) among the non-immune Uganda natives. Another typical example is that of the disease Nagana of various domestic animals in the African fly-belts, caused by the bite of a Tsetse fly (*Glossina*) which has become infected through feeding on the blood of local antelopes in which the particular trypanosome (*T. brucii*) is a normal parasite.

CONCLUSION

Our experience at Waikthlatingmayalwa in seeing the horse population of a particular district exterminated by an epidemic of disease naturally made one reflect upon the geological record which tells us how through the ages successive types of animals have completely died out. A good example is afforded by the history of the horse in the New World. Creatures which judging from their fossil bones were exceedingly like the horses of to-day abounded in the New World in prehistoric times but these had completely disappeared before man made his appearance. That this was not due to any climatic or other change making the American continent unsuitable for horses is proved by the fact that horses introduced by man from the Old World flourished and multiplied exceedingly so that vast herds of wild horses came to inhabit the prairies and pampas. What then was the cause of the extermination of the prehistoric horses? May it not have been epidemic disease like mal de caderas? May we not go further and suspect that pathogenic microbes may have played an important part in the disappearance of much more complicated creatures in the course of evolutionary history? May it not be that the final extermination of man himself when it comes will be brought about by some lowly pathogenic microbe rather than by competition with more highly developed creatures?

25th May: We left Waikthlatingmayalwa. As soon as there was daylight I got several Indians and we tackled the loalach pool and managed to get hold of six of the Lepidosirens which I promptly bundled into Kerosene tins, one in each, filling up the tin with mud in the hope that they might continue to make their etsasas *en route* or at least reach England alive.

After coffee we loaded up the cart, Budgett and I carrying out the valuables in our own hands. When all was done I distributed my surplus stores among the Indians—the young women in particular being greatly excited over a quantity of surplus dyes—bright red, carmine, and methylene blue. The cart started off immediately after breakfast while Budgett and I remained behind for a couple of hours and then rode off, waving our final farewells to our friends of Waikthlatingmayalwa.

Jogging along easily we overtook the carts, stopped near the old station and there we remained for the night, undisturbed except that the Indians heard the stealthy movements of a jaguar in the grass nearby. I took my rifle to reconnoitre but nothing happened.

Travel through the Chaco in the dry season was very different from what it had been on our previous journeys and we reached Carayá Vuelta after a journey of a little over three days.

30th May: Sibeth, myself and a couple of Indians started downstream about eight o'clock, the Indians rowing, Sibeth steering and myself sitting among the loalach boxes. There was a considerable sea running and a bitterly cold wind soon made us feel frozen right through. By midnight we were about half-way so decided to pause and make a fire to warm us up a bit. We ran the canoe against the precipitous western bank up which we managed to clamber with the aid of tree roots. On the top we found extremely dense undergrowth in which we proceeded to hack out a space of six or eight feet each way. Everything was soaking wet and it was only after much expenditure of lung power that we had at last a fine blaze and were able to heat water to make a mate drink. We stopped by the fire a couple of hours thawing and chatting and then clambered down into the canoe, I now taking my turn to steer for the rest of the way to Villa Concepción—where Budgett joined me a few days later.

6th June: Up at daylight and carted our things down to the port and embarked them on the river steamer *Aurora*. The Captain, a native of Herzegóvina, was very decent, making his men hop about in wonderful fashion, all the while dropping dark hints as to what would happen if they treated the cases roughly, seeing that they were filled with dangerous explosives. This idea was fostered throughout the trip, the cases being placed conspicuously on deck where they were regarded with interest by our fellow passengers from a safe distance. There was some mystification when a loalach poked his nose up from the mud in one of the Kerosene tins: it was pronounced to be a Boa.

In the evening after dinner I had a long chat with the Captain who gave me much information about the Balkan states and was particularly eloquent about 'los pobres Slavos' and the tyranny to which they were subjected—a new idea to myself who had been accustomed to regard the Slavs as oppressors rather than oppressed.

An uneventful journey brought us to Buenos Aires on June 13 and as our steamer for England was not due to leave until July we determined to make use of the intervening period by visiting friends further South. First we proceeded to Bahia blanca and stayed with kind friends the Wrights at their estancia. The weather was frightfully cold, with snow on the ground, and coming as we had done fresh from the tropics both Budgett and

I felt it acutely. However, the coldness of the climate was more than made up for by the warmth of the hospitality of our kind hosts. There was one event of zoological interest, in that I obtained one evening a specimen of the Pichy ciego—*Chlamydophorus truncatus*—a charming little white and nearly blind armadillo, here some 500 miles from its known haunts in the neighbourhood of Mendoza.

Early in the morning of June 25 we departed by train for Loberia and on arrival there chartered a volante for a seven league drive to the Estancia Palomar de Cobo where there resided as manager Arthur Currie, an old friend of Mate Grande days. I had warned Currie of my impending expedition to South America and had impressed upon him to do what he could to collect for me as many bones as he could of the great Edentate mammals—Glyptodons, Megatheriums and so on—which in former days roamed over the plains of South America and whose remains buried in the deposits of the pampa were often laid bare by river erosion. As Currie lived close to the Rio Quequen I had every hope that he would have some palaeosteological specimens to hand over. After a chilly drive we arrived at Currie's chilly abode and were received with the chilly, gruff manner I knew of old. He told us of the lonely life he led but recounted how some weeks earlier he had seen a stranger riding across the estancia, how he had shouted to him without any response, how he had then fired several shots at him but that he had still paid no attention: so that was that.

We spent three instructive days with Don Arturo, two of them exploring the eroded banks of the Rio Quequen grande and finding a number of bones of *Glyptodon* and *Megatherium*. The Glyptodons whose bodies were covered by a strong bony roof or carapace, superficially like that of a tortoise, interested me by the fact that when *in situ* they were always upside down; which seemed to indicate that the carcass of the creature had been floating back downwards before becoming embedded. This in turn seemed a clear indication that these pampean formations were deposited from water and that all this region had been in these ancient days an immense estuary.

Our last day at the Estancia was devoted to packing the fossil bones. We got them safely to Buenos Aires on 29th June and that concluded our work in South America—except for the lamentable discovery that the six Lepidosirens which I had hoped to transport alive to England had succumbed to the spell of quite abnormally cold weather in Buenos Aires.

CHAPTER XII

EPILOGUE

The Lepidosiren expedition, which had as its immediate object the advancement of our knowledge of the South American lungfish was, thanks in great part to a combination of good fortune with the co-operation of loyal and able helpers, ranging from red-skinned Indians to highly trained laboratory researchers, fruitful in results. From being one of the least known to science among the important types of vertebrates, *Lepidosiren* became one of the best known—more especially as regards the intricate details of its life history and the light they shed upon puzzling problems in the evolutionary history of the vertebrates.

Lepidosiren itself was, as mentioned in Chapter IX, merely the first of a trio of primitive creatures included in my original programme—the other two being *Protopterus*, a lungfish resembling *Lepidosiren* in many of its features, and *Polypterus*, an ancient type of ganoid fish, both of which linger on in the fresh waters of tropical Africa. As it happened, the Chaco Expedition provided a mass of embryological and other material of *Lepidosiren*—sufficient for a whole lifetime of research—and Budgett agreed to my suggestion that he should take over the remaining part of the programme. In fulfilment of this he made four separate expeditions to tropical Africa. The first two (1898–99 and 1900) were to the River Gambia in West Africa and on the second of these he succeeded in obtaining developmental material of *Protopterus*, but finding that it agreed closely with *Lepidosiren* in the details of its embryology, he generously handed over all this material to me. *Polypterus* had entirely eluded him, so far as eggs and early stages were concerned, so in 1902 he made a third expedition, this time to the other side of the continent, and found good localities for *Polypterus* in the upper waters of the Nile; but never any signs of spawn. In a few females he found eggs in the oviduct, and in one male he found ripe sperms but unfortunately at a time when there were no eggs available to fertilise. So this expedition, like its two predecessors, was a failure so far as regarded its main quest. In spite of all this discouragement, Budgett insisted on having another try, and this time (1903) he proceeded to the Niger delta; and now at last he succeeded in obtaining at the same time ripe eggs and sperms with which he effected artificial fertilisation. He was thus able to preserve some 200 eggs and embryos illustrating all the main stages in the development of *Polypterus*. He transported them safely to Cambridge and settled down to their study. On 9th January 1904 he had completed the first stage of his work—the preparation of drawings to illustrate the external features of the various stages and was about to commence their investigation in detail. That very evening however he developed symptoms of blackwater fever and ten days later he died. This heartbreaking tragedy, which meant an irreparable loss to the science of Zoology, imposed on me the responsibility of carrying on the investigation of the *Polypterus* material. The results, together with reprints of Budgett's published papers, and a sketch of his all too short life by Shipley, are to be found in the beautiful volume entitled *The Work of John Samuel Budgett* (Cambridge, 1907).

Although, as has been said, Budgett's first three expeditions to Africa had been unsuccessful as regards his main quest, it must not be supposed they were devoid of result. On the contrary, they had provided material for a series of highly important memoirs from his pen; on the anatomy of the adult *Polypterus* and of a 30 mm. larva, on the breeding habits of a number of West African fishes—including an account of the external features in the embryology of *Protopterus*, and on the general natural history of the regions visited.

Still other work by other workers developed out of the original Lepidosiren expedition. One after another three members of the Zoological Staff of the University of Glasgow made expeditions to the Chaco region. E. J. Bles (1905) discovered in Paraguay a most remarkable new type of wormlike animal, which was investigated in detail by C. H. Martin and named by him *Weldonia paraguayensis*, the exact relation of which to other members of the animal kingdom remains to this day a baffling mystery. W. E. Agar (1908) carried further the investigation of *Lepidosiren*, concentrating specially on the minute details of the reproductive cells so important for the understanding of heredity. G. S. Carter (1926–27) with his companion L. C. Beadle, devoted himself more especially to studying the physical conditions in which swamp animals live, and added greatly to our scientific knowledge of these as well as of the special breathing arrangements which play so great a part in enabling such animals to circumvent the dangers of the dry season. *Lepidosiren* itself has, apart from my own work, provided the material for a series of researches by others which are of special scientific importance owing to the two peculiarities already mentioned as being possessed by this animal. One of these—the relatively enormous size of the constituent cells of its body—has greatly facilitated accurate observation of their minute detail under the highest powers of the microscope. Among researches along such lines are—Agar's most important studies of the male reproductive cells, more especially of the chromosomes, the carriers of heredity; Bryce's beautiful memoirs on the blood corpuscles and their development; and the disconcerting discoveries bearing on the continuity of the nervous system by Elliot Smith, Cameron and Frances Ballantyne. All these researches mark important additions to knowledge while the last-mentioned group embody results in flat contradiction of current dogma regarding the structure and function of the nervous system in health and disease, as well as its evolutionary history.

Then again the other peculiarity—the fact that *Lepidosiren* and *Protopterus* on the one hand and *Polypterus* on the other are persisting representatives of animal groups of great geological antiquity—gives special importance to the investigation of their structure for possible evidence regarding the evolutionary history of the various organs composing the vertebrate body. Important researches of this type are those of Agar upon the skeleton of the head, of J. Robertson upon the heart and blood vessels of *Lepidosiren*, and of A. E. Miller with its corroboration of the view that the ancient *Palaeospondylus* was a creature whose affinities were with the lungfish type rather than the lamprey.

Still other investigations following more or less directly upon the original Lepidosiren expedition are the valuable embryological researches on the Swamp-eel *Symbranchus* by M. Taylor, on various Teleostean fishes by R. Assheton, and on various species of frog by E. J. Bles. These have all found a permanent place in the literature of zoological science in the form of published memoirs, but other important contributions have been without this publicity:

such as those of Muriel Robertson on the renal organs of lung-fish and of A. E. Miller upon the fin skeleton of sharklike fishes.

A general review of the scientific results emanating directly from the Lepidosiren expedition or consequential upon it show that expedition in fact to have been justified by its contributions to scientific knowledge.

To the author of this volume it meant the inauguration of a life's work of enthralling interest as a researcher in the realm of vertebrate morphology. There are those who are stimulated in their devotion to research by lofty ideals, by the ambition to add to the sum of human knowledge. I can make no such claim. My inspiration has been on a lower plane, that of personal enjoyment in playing a game—to my mind the finest of all games—that of trying to wrest from nature the secrets which she guards from mankind within her wonderful defensive entanglements of difficulty, discomfort and danger.

The field in which I have played is that of natural history—surely the most fascinating of the sciences, dealing as it does with living creatures like ourselves, living their own lives within the environment to which they are specially adapted.

It is a science which to-day is suffering through the preponderating influence of the exact sciences which deal with things that can be weighed and measured, and the failure—often on the part of distinguished scientists themselves—to realise that the exact and precise methods which are essential in the investigation of phenomena that are in themselves exact and precise are on the other hand comparatively sterile and indeed frequently misleading when applied to phenomena which are not exact and precise; and in the world of living nature this applies to most of the things that really matter. Superficial things that are the accompaniments of life—such things as differences in dimensions or temperature, rate of movement, reaction times, the amount and chemical composition of food taken in or of waste material given out—can be measured accurately, but the real thing—life itself—ever eludes the measurer and his instruments. And it is surely the high degree of this elusiveness that constitutes one of the chief attractions that leads the naturalist onwards along his narrow path, bounded on either side by the unknown, while subjected every now and then to the temptation of branching off into some seductive by-way.

PLATES

(*a*) The Biscacha (*Lagostomus trichodactylus*) [*p.* 13]

(*Copyright Zoological Society of London*)

(*b*) Pichy Ciego (*Chlamydophorus truncatus*) [*p.* 16]

(*From the Argentine Museum of Natural Sciences, Buenos Aires*)

PLATE II

(*a*) Itapytapunta. River Paraguay bounded on left by cliffs of red sandstone, on right by the low-lying Chaco bank [*p.* 43]

(*b*) Laguna with floating camelote and in the foreground leaves of the Victoria water-lily [*p.* 50]

(*a*) The *Bolivia* at Las Juntas. Entrance to northern branch of Pilcomayo [*p.* 53]

(*b*) Above Las Juntas. Overhanging trees, with a large clump of white orchid (*Brassavola*) [*p.* 55]

PLATE IV

Collared Peccary (*Dicotyles torquatus*) [*p.* 61]

The *Bolivia* aground at Fortin Page, Dam VII—the highest point reached (12th June) [*p.* 65]

PLATE VI

Our first interview with the Natokoi. Chigmaki with bowler hat and pyjama jacket; Yordaik with waistcoat and white shirt; Chinkalrdyé squatting in centre; Midshipman Page with hand on revolver; Corporal Diaz with left hand invisible through sudden movement [p. 71]

(*a*) Chigmaki [*p*. 77]

(*b*) A Guayakí Indian—one of the race of dwarfs inhabiting forests in the interior of Paraguay [*pp*. 116, 126]

PLATE VIII

(*a*) *Xiphorhynchus lafresnayanus.* $\times\frac{1}{3}$ [*p.* 93]

(*b*) Poison fangs of (left) *Bothrodon pridii* and (right) *Crotalus terrificus* (rattle-snake). Both natural size [*pp.* 103, 104]

(*a*) In the Palmar [*p*. 83]

(*b*) In the Monte—Undergrowth of Caraguatá-i [*p*. 112]

PLATE X

(*a*) (*b*)

(*a*) and (*b*) Chinerataloi, a typical Natokoi Indian.
(Bow shows backstring) [*p*. 117]

(*a*) Natokoi Indian, wearing large garment of wool, and hat made of strips of palm leaf [*p*. 117]

(*b*) A Natokoi Indian with the belt of hide worn for fighting. (The underlying woven garment being then discarded) [*p*. 117]

PLATE XII

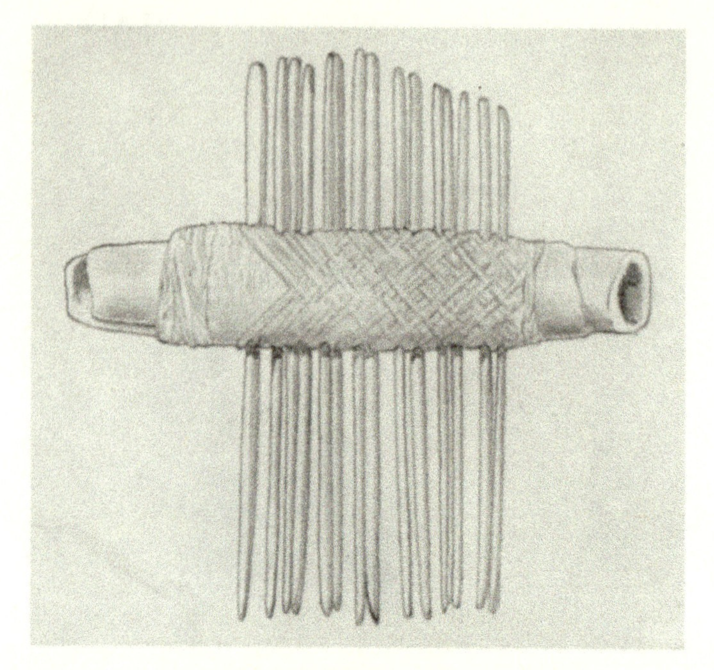

(*a*) Natokoi comb [*p.* 118]

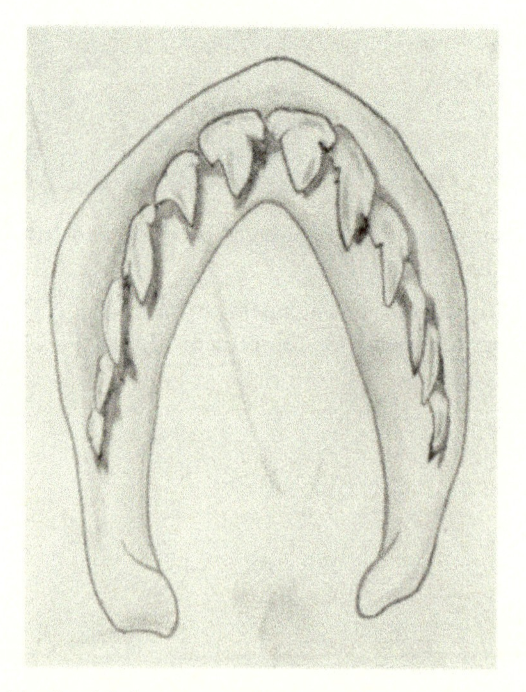

(*b*) Teeth of Palometa, used to cut hair. ×2 [*p.* 118]

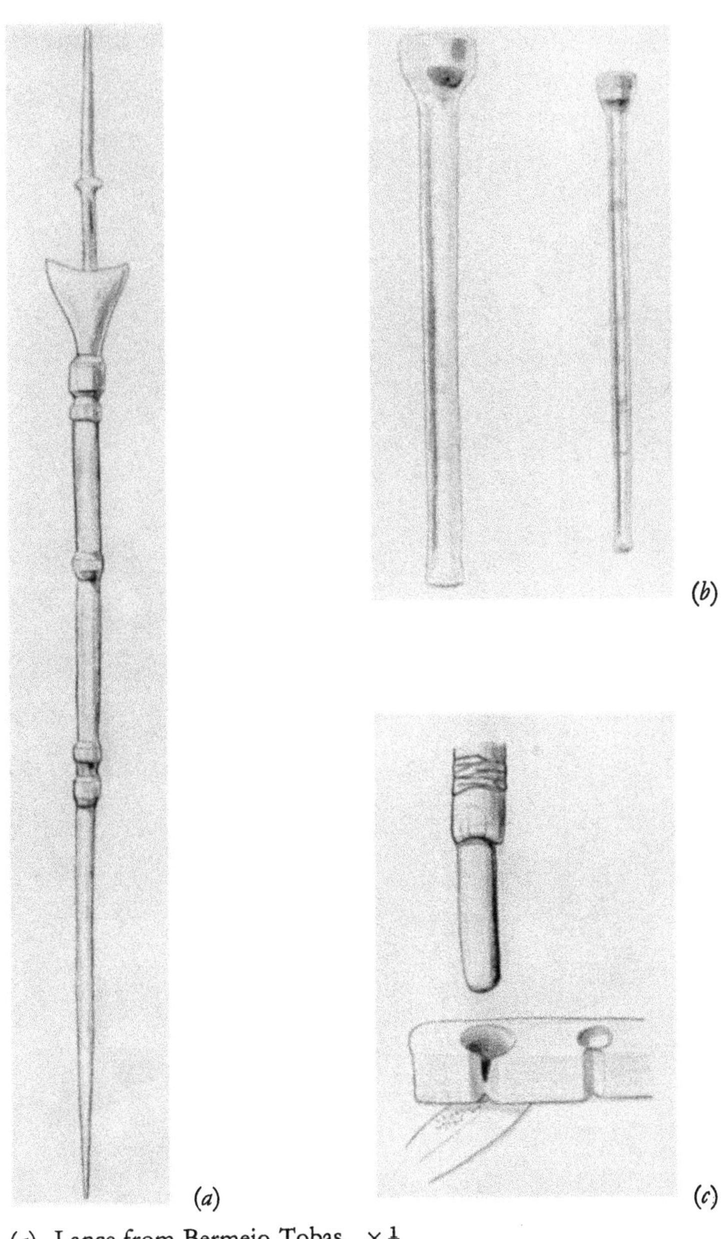

(*a*) Lance from Bermejo Tobas. $\times \frac{1}{8}$

(*b*) Club of (i) Natokoi, (ii) Mushcui, to same scale. (Actual length 2 feet $2\frac{3}{8}$ inches and 1 foot $10\frac{1}{2}$ inches respectively)

(*c*) Fire-drill. $\times \frac{1}{2}$ [*p.* 121]

PLATE XIV

(*b*) Chimaki. (San Martin) [*p*. 117]

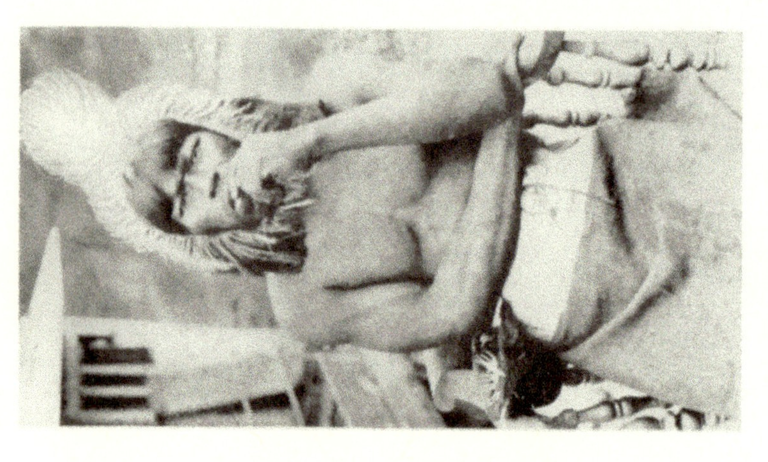

(*a*) Chimaki, wearing his tin belt and decorations [*p*. 117]

(*a*)

(*b*)

Chinkalrdyé. Note change of expression, (*a*) at ease, (*b*) alarmed and suspicious [*pp.* 77, 124]

PLATE XVI

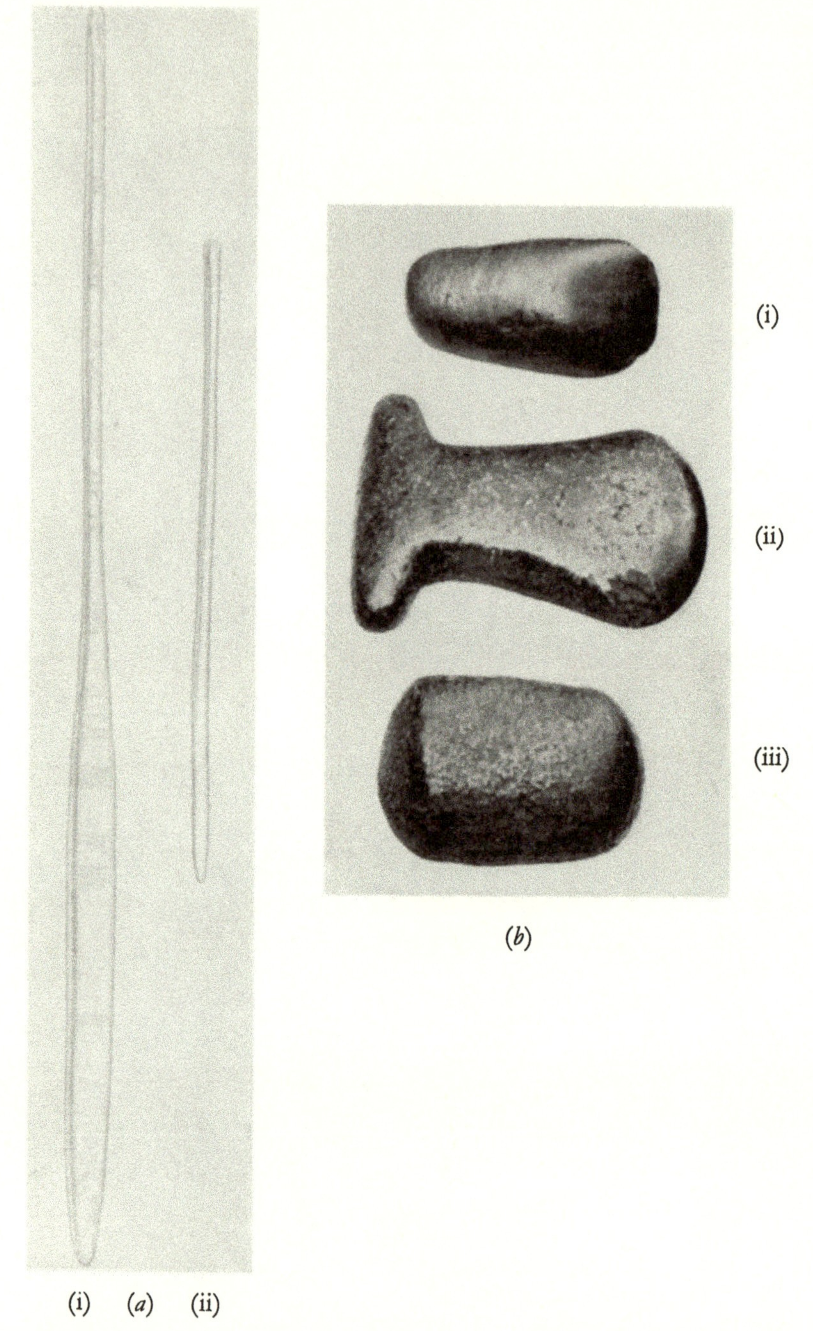

(*a*) Wooden digging implements of (i) Natokoi, (ii) Mushcui, to same scale. Total length 6 feet 4½ inches and 3 feet 3¾ inches respectively [*pp.* 125, 126]

(*b*) Three stone implements obtained from the stomach of *Rhea*. (i) and (ii) axe-heads; (iii) hammer. (Reduced by $\frac{1}{10}$) [*p.* 126]

Group taken on arrival in Asunción by San Martin. Graham Kerr, Chimaki and H.B.M. Consul Dr William Stewart [*p.* 157]

PLATE XVIII

My last view of the *Bolivia* [*p.* 163]

PLATE XIX

(*a*) The three surviving types of lungfish. A, *Ceratodus* (Queensland); B, *Protopterus* (Africa); C, *Lepidosiren* (South America) [*p.* 170]

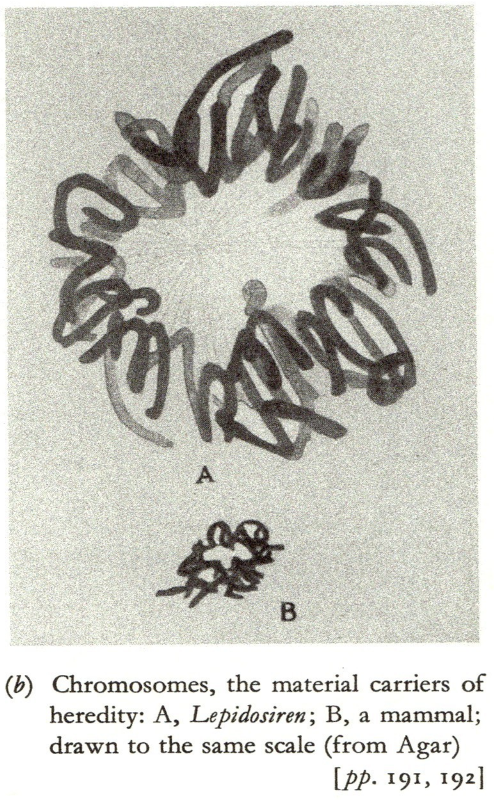

(*b*) Chromosomes, the material carriers of heredity: A, *Lepidosiren*; B, a mammal; drawn to the same scale (from Agar) [*pp.* 191, 192]

PLATE XX

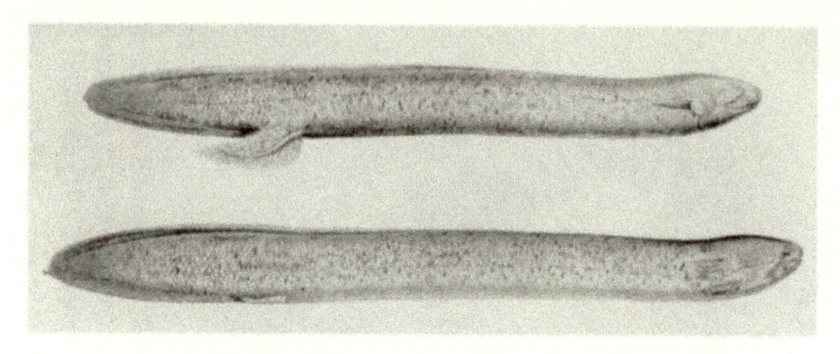

(*a*) *Lepidosiren paradoxa.* Male (above) during the breeding season and female [*p.* 180]

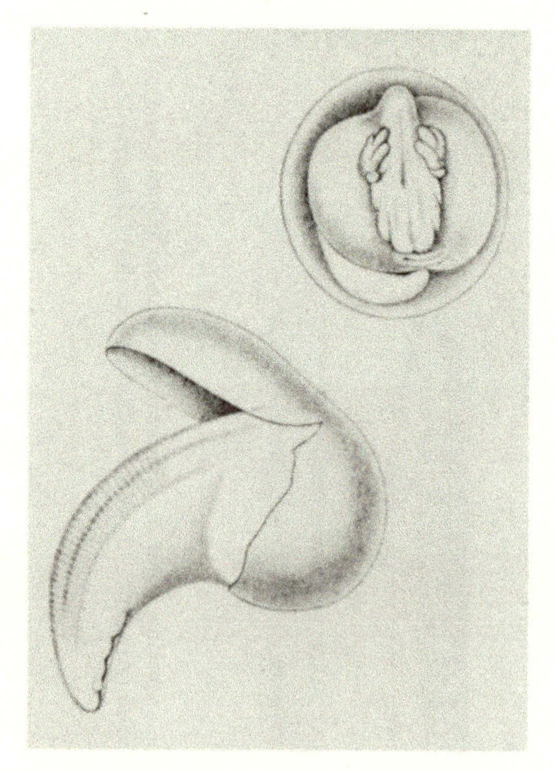

(*b*) Egg of *Lepidosiren* (above) just before hatching; (below) during hatching. ×3 [*p.* 184]

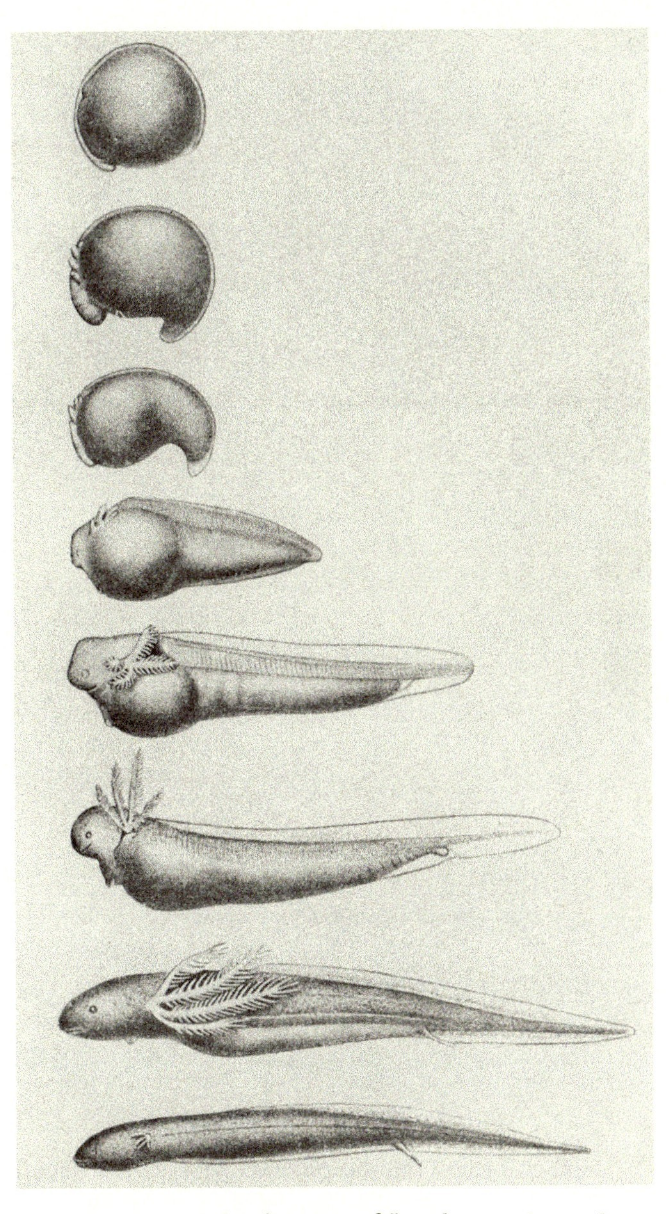

Stages in the development of *Lepidosiren* [*p.* 184]

PLATE XXII

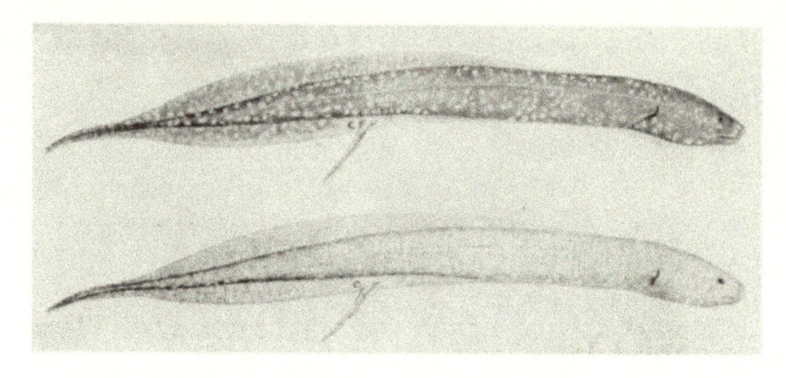

(*a*) Day and night appearance of young Lepidosiren [*p.* 189]

(*b*) A Lengua (Mushcui) canoe on the River Paraguay, near Concepción [*p.* 176]

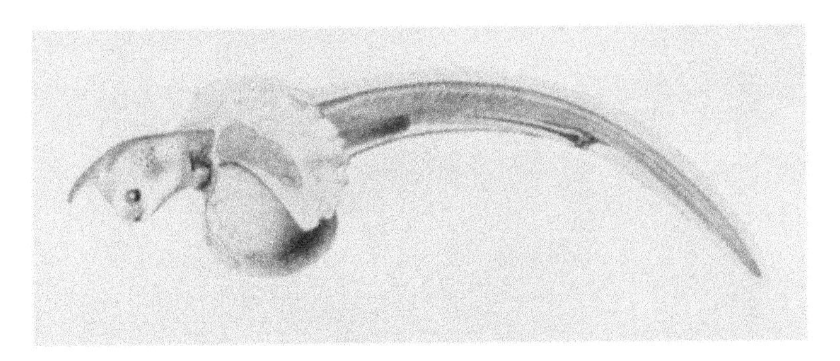

(*a*) Larva of *Symbranchus*. ×7 (from Taylor) [*p*. 192]

(*b*) Transport of one of the Lepidosiren packages through swamp stream [*p*. 205]

PLATE XXIV

Group taken by Andrew Pride at Waikthlatingmayalwa. Europeans from left in middle row: Wilhelm (cook), John Hay, W. Barbrooke Grubb (missionaries), Graham Kerr, Budgett, R. J. Hunt (missionary). Europeans in back row: Insley, Sibeth (stockman), Graham, Hawtrey, Mark (missionaries) [p. 221]

INDEX

www.ingramcontent.com/pod-product-compliance
Ingram Content Group UK Ltd.
Pitfield, Milton Keynes, MK11 3LW, UK
UKHW040657180125
453697UK00010B/241